The Victoria Cross

The Victoria Cross, Great Britain's highest award for bravery, was instituted by Queen Victoria in 1856, many believe at the suggestion of Albert, The Prince Consort. The original warrant, of 29 January 1856, stated that '...the Cross shall only be awarded to those officers and men who have served Us in the presence of the enemy, and shall have then performed some signal act of valour or devotion to their country.'

The Victoria Cross itself is a bronze, straight-edged cross patée. It measures 1½" across, with raised edges. On the obverse side, in the center of the cross, is a Lion Guardant standing upon the Royal Crown. Below the Crown, in a semi-circular scroll, are the words 'For Valour.' On the reverse of the cross the date of the award is engraved in a circle in the center.

At its inception the Cross was carried on a red ribbon for army personnel and blue for the navy, but in 1920 this was amended and all subsequent crosses were hung from a plain red ribbon.

Traditionally, the Victoria Cross was manufactured from the bronze of guns captured from the Russians in the Crimean War of 1854–1856, but this supply was exhausted in 1942.

In the First World War of 1914–1918, the Victoria Cross was won by 19 airmen while serving in the Royal Flying Corps, The Royal Air Force, and the Royal Naval Air Service.

Name and Rank	Date of Award	
Lieutenant W.B. Rhodes-Moorhouse †	22 May	1915
Flight Sub-Lt. R.A.J. Warneford	11 June	1915
Major L.G. Hawker, DSO	24 August	1915
Captain A.J. Liddell, MC	28 August	1915
Lieutenant G.S.M. Insall, MC	23 December	1915
Lieutenant R.B. Davies, DSO, AFC	1 January	1916
Major L.W.B. Rees, OBE, MC, AFC	5 August	1916
Captain W. Leefe Robinson	5 September	1916
Sergeant T. Mottershead, DCM †	12 February	1917
Lieutenant F.H. McNamara	8 June	1917
Captain A. Ball, DSO+2Bars, MC, CC, L.d'H. †	8 June	1917
Lt. Colonel W.A. Bishop, DSO+Bar, MC, DFC, CC, L.d'H., C.de.G.	11 August	1917
Major J.T.B. McCudden, DSO+Bar, MC+Bar, MM, C.de G.	2 April	1918
Lieutenant A.A. McLeod	1 May	1918
Lieutenant A. Jerrard	1 May	1918
Captain F.M.F. West, MC	8 November	1918
Captain A.W. Beauchamp-Proctor, DSO, MC+Bar, DFC	30 November	1918
Major W.G. Barker, DSO, MC+2Bars, CC, L.d'H, C.de G., VM	30 November	1918
Major E. Mannock, DSO+2Bars, MC+Bar †	18 July	1919

† = posthumous award

Introduction

The First World War was the first war in which airplanes were used extensively. This very fact, that one aspect of the war of 1914–1918 was fought in the air, made acts of bravery, unprecedented in the new element, seem even more courageous and worthy of merit. This would partially explain the award of the Victoria Cross in the early years of the conflict for acts which would become almost commonplace in the later years. This is not to detract from those first acts of bravery in air combat which were rewarded with the VC. Men fresh from the ways of peace, unused to the horrors of war, without a thousand precedents, perhaps showed more courage than those who came later, hardened in the crucible of war.

Again, acts of valour in the air which were the first of their kind were rewarded with a VC. On 24 September 1916, only three weeks after Robinson's destruction of the airship SL11, for which he deservedly received a VC, Lt. F. Sowery shot down LZ74. Nine days later Lt. W.J. Tempest destroyed LZ72. Both Sowery and Tempest were no less brave than Robinson, and both were awarded the DSO, no mean honour in itself, but not the VC. Robinson was fortunate in that he was the first airman to destroy an airship over England.

There are two instances of single actions which resulted in the award of the VC which have caused controversy over the subsequent years. Bishop – whose award was exceptional in that it is the only VC to be awarded on the word of the participant alone, without confirmation or witnesses of the action for which it was given – and Jerrard, who seems to have been the unknowing victim of an over enthusiastic combat report by the commanding officer of his squadron. The historical writer is faced with a dilemma in such cases. Should he gloss over the findings of recent research and compound the errors and distortions of the past; or should he present the known facts and risk giving offense to interested parties. It has been written that the first and best purpose of historical writing is to supplant long-held and widespread distortions with informed insight. This I have tried to do.

These reflections are not intended to devalue in any way the single acts of bravery which led to the award of the VC, but in the final analysis perhaps the most valid of the VC awards were not those given for one act of superb bravery – for many similar acts must have passed almost unremarked – but for courage sustained over a long period. Few would deny that such courage and devotion to duty shown by men such as Ball, Mannock, McCudden, and Beauchamp-Proctor are remarkable in the annals of the Royal Flying Corps and the Royal Air Force in the Great War of 1914–1918.

Alex Revell
Nancledra
Cornwall, 1997

Captain Albert Ball

"The first of the great British fighter aces was Albert Ball, and his name flashed through the history of the first war with the intensity of a meteor." wrote Marshal of the Royal Air Force, Lord Douglas of Kirtleside, who, as Major Sholto Douglas, was a young squadron commander in 1917.

Albert Ball was born in Nottingham on 14 August 1896, the first son of Albert and Harriet Ball. Albert Ball senior was a successful business man with a keen sense of civic duty; he entered local politics in the closing years of the nineteenth century, and in 1899 was elected a city councilor of Nottingham. Over the next ten years he was a Justice of the Peace, an Alderman on the City Council and, in 1909, was elected Mayor of Nottingham, an office he was to hold four times during his lifetime.

The family – a daughter Lois, born in 1892, Albert's younger brother Cyril, born in 1898, and Albert – had moved to Sedgley House, in the Park district of Nottingham, soon after the birth of Lois, and it was here that the future VC spent his childhood. It was a secure, comfortable boyhood. Albert's father was very supportive of all his children and in particular he encouraged his elder son's interest in all things mechanical. A small wooden shed was erected behind Sedgley House and Ball spent nearly all his time there; building wireless and telegraphy equipment, and experimenting with various mechanical devices and inventions. From the age of nine, Ball began to collect pistols, firing them at sticks stuck in the ground of the tennis court in the garden of Sedgley House, and he soon became an accurate shot. On one occasion, when he was about fourteen years old, he arrived home, followed by a band of admiring boys, pushing a wheelbarrow containing a derelict engine. Ball stripped down the engine in his hut and soon had it running again.

In January 1911, Albert and Cyril were enrolled in Trent College, Long Eaton. The college was run by a strict disciplinarian and the Ball boys, used to their indulgent father, found the rigorous regime at Trent something of a shock. However, Ball was reasonably happy at the college; although not particularly academic he found much to interest him: he became a competent draughtsman, took up carpentry and photography, and began to study the violin.

Ball left Trent College in 1913. He was a small but stocky seventeen year old, eager to make his way in the world. His father found him a position with a small engineering works in Nottingham and Ball was happy to experiment and learn in his first employment. But it was to be a short lived period of happiness. On the declaration of war in August 1914, Albert Ball was one of the many thousands of eighteen year old youths who volunteered to fight for King and Country, and on 21 September he was enlisted as a private in the Nottingham and Derbyshire Regiment – colloquially known as the Sherwood Foresters.

In June 1915, Ball was attending an officers' training course at Perivale in North West London, and it was here, only a short distance from Hendon Aerodrome, that Ball first conceived the idea that he could learn to fly and subsequently join the Royal Flying Corps. He enrolled in the Ruffy-Baumann School at Hendon for a course of flying lessons, which meant that he had to leave his billet at 3:30a.m., ride his Harley-Davidson motorcycle to Hendon, have his lesson at four o'clock, in the calm morning air, and return to Perivale in time for the first parade of the day at 6:45a.m. But to Ball it was worth the effort. He wrote home: 'I think I shall make a good pilot.'

On 15 October 1915, Ball took the test for a Royal Aero Club Certificate, flying a Cauldron. He passed easily and was awarded Certificate No.1898. He was then accepted into the RFC on a probationary basis and reported to 9 Reserve Training Squadron in Norfolk. On arriving in Norfolk, Ball was given the chance to go to France within the next few weeks, providing that he gave up the idea of being a pilot and trained as an observer. The newly-formed 18 Squadron, equipped with the Vickers FB5 – the Gunbus – a two-seater pusher fighter, was short of observers and was about to proceed to France. The temptation was there, but Ball resisted it. He had paid good money to learn to fly and his business instinct deplored the waste of time and money if he re-mustered as an observer.

On 19 December, Ball was posted to the Central Flying School at Upavon in Wiltshire to begin his service flying training. Despite the time of the year, with its attendant appalling weather, Ball passed out from the CFS in just over a month, winning his 'wings' on 29 January 1916, and was finally transferred to the RFC. However, Ball's hopes that he would now be posted to a squadron on active duty in France were dashed by his next orders, received the following day. He was posted as a flying instructor to 22 Squadron, stationed at Gosport, Hampshire.

Although disappointed at not being sent to France, it was not in Ball's nature to not give his best at any job he was given. During his time at Gosport, flying Avro 504s and Martinsydes, he worked extremely hard. He was still very much an embryo pilot himself, but nevertheless, on one day alone, in January 1916, he took up over 30 officers for instruction, six of whom he sent off on their first solo flight. On 15 February his patience was finally rewarded. Albert Ball wrote a hurried note to his family – his 'dearest people' – at Dover on 17 February 1916. 'It is strange to be leaving your home on such a mad job, but, cheeroh you dears, I shall soon be back. Good bye.'

Ball's posting was to 13 Squadron, based at Vert Galant, France. The Squadron was equipped with BE2c and BE2d two-seater aeroplanes and its duties were general reconnaissance, photographic, and ranging for the artillery. Ball had arrived at a critical time in the fortunes of the RFC. Since the early summer of 1915 a new shape had appeared in the skies over France: the Fokker E-type monoplane. Although not particularly fast, nor a very sturdy aeroplane, the Fokker *Eindecker* had one great advantage over its opponents: it was armed with a

Albert Ball.

fixed LMG08 machine gun firing through the arc of the propeller, enabling the aeroplane to be aimed like a gun along its line of flight. Armed with the LMG08, capable initially of firing 400 rounds a minute – this was later raised to 600 – the Fokker monoplane, in the hands of the emerging German fighter pilots, decimated the slow, unmanoeuvrable reconnaissance aeroplanes of the RFC. By the end of 1915, the Fokkers, until then distributed in ones and twos among the two-seater units of the German Air Force, were reformed into groups. These non-permanent groups, *Kampfeinsitzerkommandos,* were deployed tactically along the Front and the Fokker pilots could now hunt together. The RFC was forced to adopt the defensive tactic of flying in formation in an attempt to meet the threat of the Fokkers. RFC HQ orders were that every aeroplane on reconnaissance or photographic duty had to be escorted by at least 'three other fighting machines.' All the machines were to fly in close formation and if any became detached the mission should be aborted. The Fokkers had effectively cut the strength of the RFC by two thirds.

On 21 February 1916, Ball flew his first patrol in France. The previous day the squadron had lost a crew to the Fokkers. 2Lt. F Garlick and Captain W. Knox had both been killed in action in BE2c 2054 while on patrol over Aizecourt-le-Bus, claimed by *Ltn.* Leffers of *KEK* Bertincourt.

Ball, with Lieutenant Green as his observer, took off with five other BEs on the morning of 21 February and made for their designated patrol area. The formation came under heavy and accurate anti-aircraft fire – known popularly as 'Archie' by the RFC – and were attacked by three Fokkers and an assortment of other German aircraft. In accordance with their orders, the BEs broke off the combat and retreated to the safety of the British front line, but Ball's engine gave trouble and he was forced to land just behind the British trenches. After spending the night with the infantry, Ball took off again the following morning. Despite a snowstorm, which forced him to land

The Ball boys, Cyril (left) and Albert.

several times, he used his skill with engines to coax the failing engine of the BE to run well enough to return to his base, landing at eight o'clock. This episode had a fortunate outcome for Ball. When he had arrived in France, his commanding officer, Major Marsh, had considered him a poor pilot – despite his having been an instructor at Gosport – and had decided that he should be sent back to England for more instruction. Ball's bringing back the BE from the forced landing convinced Marsh of his worth.

Over the next weeks Ball flew the normal patrols undertaken by the Squadron. These resulted in no combats, but were not without incident. Taking off on 20 March, the engine of BE 4352 cut out and it was wrecked in the ensuing forced landing, trapping Ball and his observer, Lt. S.A. Villiers, in their cockpits. Luckily there was no fire and Ball and his companion were pulled from the wreckage.

It was not until 30 March that Ball had his first combat with a hostile aeroplane. Again flying with his friend Villiers – who the morning before had been in combat with two Albatroses over Vimy, he and his pilot barely making it back to base in a badly damaged BE – Ball was artillery spotting between Vimy and Givenchy when he sighted a German two-seater below and dived to attack it. As Villiers opened fire with his Lewis gun at the fleeing two-seater, there was the sound of machine gun fire from behind them. A second enemy aircraft had dived onto the tail of the BE and its fire cut a drag wire. Ball turned to meet the new threat, but the German pilot continued his dive and escaped. This incident was an almost exact parallel to Villiers' escape of the previous day, in the same area, and it is probable that it was a prearranged plan by the enemy pilots.

April saw a gradual improvement in luck for Ball and Villiers. On 10 April, returning from escort duty, they attacked an observation balloon and caused it to be hastily hauled down. Flying an artillery shoot on 29 April, Ball, again flying with Villiers, was attacked by five Albatros C.III two-seaters. Undeterred by the odds, Ball turned towards the enemy machines, but the enemy formation turned away and dispersed, only one of the Albatros pilots staying to fight the BE. Villiers had fired only five shots at the attacker before the enemy observer was seen to slump down in his cockpit. The Albatros pilot immediately dived away and forced landed south east of Rouvroy. Ball and Villiers had been so close to the enemy machine that Ball later wrote home: 'we could see the Huns' faces and they could see ours, we were so near.'

At this time some of the RFC Corps squadrons had one or two single-seaters on charge, flown by pilots who were thought to show the right quality of aggressive spirit – the first fighter pilots of the RFC. Pilots were encouraged to try their hand on the little single-seaters – 'scouts' – as they were known – and Ball was one such. On 23 April he had been sent to the No.1 Aeroplane Depot at St.Omer to learn to fly the Morane Bullet. The little Morane was a notoriously difficult aeroplane to fly – one pilot later said that it was the only aeroplane he ever flew, during a long service career, that he was convinced was actively trying to kill him – but Ball managed to fly it without incident. On his return to the Squadron he constantly requested Major Marsh to let him fly the Squadron's Bristol Scout Type D, and after the successful fight on 23 April, Marsh finally gave his permission, Ball taking off after lunch for a 30 minute flight in the scout. Ball flew only in the vicinity of the aerodrome but was overjoyed with the speed, lightness of control, and manoeuvrability of the little Bristol after the slow and stable BE2c. That evening, he took off again in the Bristol and made for the front line, hoping for a combat, but he saw no enemy aircraft.

Another pilot crashed the Bristol a few days later, but a replacement was on charge by 5 May and Ball took off in the evening to look for enemy aircraft. Crossing the lines he decided to test and warm his gun by firing a few rounds. It was fortunate he did so. The early interrupter gear was faulty and the bullets hit the propeller of the Bristol, almost sawing it in half. Ball hastily returned. As he wrote in a letter home the following day, 'If I had fired about six more shots I should have had to land in Hun land, for I could never have got back.'

Two days later, after flying a morning patrol, his last in 13 Squadron, Ball walked across the aerodrome at Savy to the hangars of 11 Squadron. He was carrying a note from Major Marsh to Major T.O.B. 'Mother' Hubbard, commanding No.11 Squadron, saying he could be trusted

Ball flew this Nieuport 16, serial A126, while in 11 Squadron.

to fly 11 Squadron's Nieuport Scout. It was 7 May 1916, the beginning of Ball's career as a fighter pilot. He had exactly one year to live.

In 1915, 11 Squadron was the first squadron to go to France with the Vickers FB5. When Albert Ball joined squadron strength, it was an experienced unit, one of its members having already won a Victoria Cross. One of the first people Ball met on the morning of 7 May was 2Lt. A.J. Insall, the brother of Gerald Insall VC. Ball handed Insall the note from Marsh. 'He is a keen conscientious young man, and should do well,' it read. Insall respected the judgment of Marsh, who he knew well, but his first impression of this new pilot was not altogether favourable and he placed Ball in a Flight other than his own. 'He seemed a bright young fellow, inclined to rather more self-confidence than the run of pilots with his flying hours, and I put him down for one of the two other Flights.' It was a first impression that was later to change.

The first thing Ball did on joining 11 Squadron was the mundane task of seeing to his material comforts while off duty. Not liking the billet found for him in the nearby village of Aubigny, he chose instead to pitch his tent on the aerodrome. This had the effect of placing him near his aeroplane, ready to take off if any enemy machine was reported over the lines. A space of some 40 square yards was marked out around the tent by wire, and Ball began to cultivate a small garden, where he could spend his evenings. He wrote home. 'My work is rather a nerve pull, so I think it best for me to forget all about it after it is over.' Ball's reference to his nerves was justified. Although he had been in France for only 11 weeks, he had flown over 40 operational missions, with many additional hours flown in testing. Work in the garden was an invaluable method of relaxation. The garden, with the aid of seeds sent from home, soon flourished and the tent was replaced by a small wooden hut.

Bad weather stopped flying for some days after Ball's arrival at 11 Squadron and it was not until the morning of 16 May that he had his first combat in his new squadron. Taking off in the morning, in Bristol Scout 5312, he was at 12,000 feet over Givenchy when he saw an Albatros two-seater 7,000 feet below him. Ball dived on the enemy two-seater and fired 120 shots. The Albatros turned upside down at 2,000 feet and fell out of control, but Ball was unable to follow it down to see its fate. His engine had begun to miss, possibly the result of his prolonged, steep dive, and he was forced to pull away. Nevertheless, this Albatros was awarded to Ball as his first victory. Ball's jubilation, however, was tempered by events later in the day. One of his dearest friends in 13 Squadron, Captain Lucas, was shot down and killed that evening by a Fokker *Eindecker* over Anzin St.Aubin. Captain G.B. Lucas was hit in the Fokker's first burst of fire and his pilot, Lieutenant Wright, although hit in the shoulder himself, turned back and managed to land the badly damaged BE behind the British Lines. Ball wrote home the next day: 'Lieutenant Wright...walked out a few yards and then fell down. Captain Lucas died in a few hours. The machine was brought back this morning and I am not exaggerating when I say it was soaked in blood and full of bullet holes.'

Ball's own squadron had also not been without loss on 16 May. 2Lt. M.M. Mowatt, flying 5301, the other Bristol Scout on charge, was shot down near Izel at 5:30p.m. by *Oberleutnant* Max Immelmann of *Feldfleigerabteilung* 62, flying a Fokker. Mowatt was the 15th victory of the German ace.

Three Nieuport 16s were on 11 Squadron charge at this time: A117, A126, and 5173. Ball damaged the lower right wingtip of A126 in a bad landing on 27 May and it was returned to 2AD for repair. Ball was then given the use of 5173 and, after his victory on the Bristol Scout on 16 May, Ball almost exclusively flew this Nieuport for the next weeks, having his first combats in it on 22 May: three inconclusive fights, first with two separate Albatros two-seaters, then an LVG.

On 29 May, Ball scored his first victory in the little Nieuport. Taking off in the morning, Ball climbed to 8,000 feet and crossed the front line south of Arras. At 7:30a.m. he spotted an Albatros two-seater at 7,000 feet over Boiry, closed the range, and opened fire, fighting the enemy machine down to 3,000 feet and expending a full drum of Lewis gun ammunition. Ball was having difficulty in changing his drum and turned away. The Albatros pilot took the opportunity to escape and dived under the cover of his own batteries around Douai. Ball flew on. Ten minutes later saw another Albatros over Moyenneville, south of Arras. The enemy machine was at

The cockpit of Ball's Nieuport 16.

6,000 feet and Ball closed to within 30 yards range before firing half a drum of Lewis into it. The Albatros banked steeply then dived vertically. Ball lost sight of it in the ground mist and was unable to see whether or not it crashed. Ball then flew towards Douai and 25 minutes later saw another Albatros, this one escorted by two Fokker E types. Ball kept his distance while he waited for an opportunity to attack the Albatros on its own. But two additional Fokkers then appeared, flying above the others, and Ball decided to look elsewhere for an easier target. Just as he started to bank away, the enemy two-seater turned towards Oppy, its escort leaving it and flying towards Douai. Ball took his chance and attacked the Albatros over the trenches by Oppy, closing to 50 yards and firing the remaining half a drum of Lewis into it. The enemy machine dived away, the observer still shooting at the Nieuport, putting eight shots into the fuselage, one bullet hitting just behind Ball's back. Ball watched the Albatros land in open country, then turned for Savy. On the way back the tailplane of the Nieuport was hit by anti-aircraft fire. Ball was credited with the second Albatros as out of control and the third as forced to land.

Ball next scored on 1 June, a victory over a Fokker E-type. He took off in 5173, climbed to 10,000 feet, and flew to the German aerodrome at Douai, where he circled, waiting for any enemy machines to come up to attack him.

Ball waited for nearly half an hour, but eventually two enemy machines took off: an Albatros two-seater and a Fokker *Eindecker.* The Albatros reached Ball first. Ball dived and fired ten rounds at it. The Albatros pilot hurriedly dived to land back on Douai aerodrome. By this time the Fokker had arrived and was on Ball's tail, closing the range before opening fire. Ball waited until the Fokker pilot commenced firing, then turned to meet his attacker. No sooner had Ball begun to return his fire, then the Fokker pilot also turned away, dived, and landed in a field two miles to the west of his aerodrome. Ball was awarded the Fokker as a forced to land victory.

Ball was in action again on 8 June. Taking off at half past five, he first attacked an Albatros over Méricourt at 6:15a.m., but was frustrated by a gun jam. Fifteen minutes later he attacked another Albatros over Acheville/Rouvroy at 7,000 feet, but the enemy pilot dived away and avoided his fire. It was another ten minutes before Ball saw another chance of a victory: an enemy machine at 7,000 feet over Dourges. Ball drove this machine down, but gained no decisive result. Twenty minutes later he had better luck, forcing an Albatros to land near Flers at 7:00a.m. Ball's last combat on this eventful morning was with an LVG over Rouvroy, driving it down from three to two thousand feet before he ran out of ammunition. Ball was awarded no victories from these combats.

Ball left France for a well deserved leave on 10 June and he returned to the Squadron on 23 June, feeling fit and fresh. It was the beginning of his most successful period.

His first victory after his leave came on 25 June. In the preparations for the forthcoming battle of the Somme, RFC HQ ordered that the German balloon line be destroyed. In the morning, Ball's first attempt at his selected balloon showed no result. He took off again in the afternoon and this time was successful, destroying the balloon with phosphorus bombs. The ground defences hit the Nieuport in many places, including the engine, and Ball had to nurse the crippled fighter for eight miles, until he reached the safety of the British front-line trenches. Not surprisingly, Ball thought destroying balloons 'a rotten job' but he was awarded a Military Cross for his work on the afternoon of 25 June.

A certain amount of re-organization was now taking place in 11 Squadron. On 27 June, Ball, along with 2Lts Griffiths and Anderson and Sergeant Reffell, were grouped into one 'scout' flight: 'B' Flight. It was an inauspicious day for Ball. He took off at 7:00a.m. in Nieuport A134; as he crossed the front lines he saw a German troop and transport concentration in a wood. Going lower for a closer look, the Nieuport was badly hit by anti-aircraft fire, one of the cylinders was smashed

Ball flew Nieuport 16 A201 in 60 Squadron, scoring 11 victories while flying it.

and a piece of shrapnel narrowly missed Ball's leg. With no engine, Ball turned back and just managed to land far enough behind the Lines to prevent the enemy from shelling the Nieuport. Ball sent half of the smashed cylinder home.

On 1 July 1916 the Battle of the Somme began. Over the next months, until winter called a halt to the offensive, Ball was to score 26 victories, and was to become, in the words of the official history 'the spearhead of the achievement of the Flying Corps over the Somme.'

Ball was in the air on the first day of the offensive, but it was not until the following afternoon that he claimed a victory. Taking off as an escort to four FE2bs, Ball crossed the Lines with the two-seaters and spotted a formation of six Roland C.IIs coming from the direction of Mercatel. The FE formation split, two circling the Front Line to carry out their duties, the other two, with Ball, attacking the enemy machines. One of the FE crews sent down one of the Rolands and Ball closed with another. A full drum of Lewis into the fuselage of the Roland sent it down to join its companion, both crashing by the side of the road from Arras to Mercatel. It was 5:30p.m. Half an hour later, Ball caught an Aviatik C-type over Lens. Ball carefully stalked the two-seater, eventually getting to 20 yards before being seen. He opened fire with his Lewis, which only fired a few shots before stopping with a jam, which Ball managed to clear. He dived below the Aviatik, pulled down the Lewis on its Foster mounting, and fired a full drum into the underside of the enemy machine from point blank range. The Aviatik went down out of control and crashed into a field.

The effectiveness of his method of attack against this Aviatik, discovered almost by accident in the heat of the combat, was not lost on Ball. From this time, many of his victories were gained by this method of attack, at point blank range from underneath, and it became his trademark.

The double victories on 2 July were awarded as only the second and third of Ball's victories which were officially confirmed. Up to this date, since joining 11 Squadron as a fighter pilot, he had had 14 fights, sending two machines down out of control and forcing four to land. By his own private reckoning, he thought that his score stood at seven, including the balloon on 25 June, but it was not in his nature to claim anything which he felt was in the least doubtful. His combat reports are scrupulously careful to claim nothing, merely detailing the possible result of each combat, without conjecture or exaggeration.

Ball's next combat was another 'rotten job': an attack on a balloon near Pelves. Ball tried a low attack, crossing the Lines at only 5,000 feet, but the balloon crew saw him coming and started to haul down the balloon. Ball closed the range to within 15 yards, loosed off his eight Le Prieur rockets, and fired a drum of Buckingham ammunition at the descending balloon. Although the rockets missed, the bullets found their mark, but the gray mass failed to ignite, and Ball made a point of this in his combat report.

Although Ball was a determined fighter, giving his all to the job in hand, he still regarded the war as an interruption to his life and commercial aspirations. Shy and reserved he may have been, but he was nevertheless impatient with military discipline and had little respect for higher authority. This attitude was to land him in trouble. The intensity with which Ball fought and approached his daily work with the squadron had left his nerves, never it seems very strong, in a dangerously stressed state. On 16 July, after ferrying one Nieuport to the depot at Candas and returning with another, he went to see Major Hubbard. He was desperately tired, and quietly asked if it would be possible to have a few days rest from his constant flying duties. 'Mother' Hubbard was a kindly man, but instead of quietly granting Ball's wish and keeping him on the ground for a few days, he passed the request on to Brigadier-General Higgins – known as 'all bum and eyeglass.' The result was that Ball found himself posted to 8 Squadron to fly BE2cs on general reconnaissance work. It may well have been that this was a result of his request being turned down by Higgins and Ball's outspoken reaction to its refusal. William Fry, a pilot who met Ball later, on Ball's eventual return to 11 Squadron, recalled in his memoirs, *Air of Battle,* that Ball was intolerant of senior officers, who lacked any knowledge of the work he was doing, and that he had been sent to 8 Squadron for being too outspoken. It was also possibly Higgins' way of disciplining Ball a little. Ball had been outspoken – especially so for a junior subaltern – to 3rd Brigade in respect of some of the aeroplanes he and his comrades were asked to fly. The story has been told that after he had landed a BE12 from a test flight, he had been asked by General Trenchard what he thought of it. Ball had replied 'It's a bloody awful machine.'

Hubbard and several other officers were appalled at Higgins' treatment of Ball, and Hubbard wrote to the commanding officer of 8 Squadron; 'You will find Ball a good little chap if managed in the right way. He is young so naturally wants a little more rope than the other pilots.' Hubbard then added a list of Ball's combats since joining his Squadron and ended the note by saying that if, after a few weeks, Ball wished to return to 11 Squadron, he would be only too pleased to have him back.

Ball was not so complacent about his posting. He wrote to his father, 'This is thanks after my work, and not even the Major can stop it. Three Majors have done their best for they all think it is a cad's trick. However, here I am. Now I think I shall ask to go back to my regiment. Will you advise me, write soon. Address to No.8 Squadron, BEF. PS. Oh, I am feeling down in the dumps.'

It was not in Ball's nature to stay depressed for long and he resolved to make the best of his enforced 'rest.' In the event it turned out to be far from that. On 21 July, returning from a bombing raid, Ball's BE2c developed engine trouble. He fell behind the rest of the formation and two Fokkers appeared. Ball later wrote 'I thought it was all up with A.B.' However, the German pilots did not attack and Ball returned safely.

Before he returned to 11 Squadron, Ball volunteered for another job he described as 'rotten' – in this case the landing of a spy behind the German lines. The spy, known as '*Monsieur* Victor' had twice been taken to his destination by a pilot of the squadron, but on each occasion a landing had not been possible due to the heavy ground mist.

Ball and '*Monsieur* Victor' left the Squadron's aerodrome at La Bellevue at 8:15 on the evening of 28 July. Evading an attack by three hostile machines, Ball flew further into German-held territory, attracting some anti-aircraft fire, but as darkness fell he arrived at his destination and landed safely. To Ball's annoyance, 'Victor' refused to leave the aircraft, pointing out, not unreasonably, that the attention the BE had attracted had warned the Germans of his presence. Ball took off and landed a further three times in an attempt to get the spy to disembark, but 'Victor' was adamant, and Ball had to take him home, arriving back at La Bellevue at ten minutes to ten.

Higgins arrived at the aerodrome the next morning, full of congratulations – a reaction not unlooked for by Ball. It was part of his plan to get back in the General's good books. But he was not to return to 11 Squadron for another two weeks. He wrote home; 'Oh, I shall be glad to get back to my Nieuport...they don't give you a rest unless you are quite a crock, or are too whonkey to do your job.'

Ball returned to 11 Squadron and his 'beloved Nieuport' on 14 August, his 20th birthday. While he had been away he had arranged for his garden to be tended by a local gardener and the little plot was still flourishing. Best of all, a brand new Nieuport 16 had been assigned to him, No.A201. He wrote home on 16 August: 'Dearest people...Hello, I am back in my dear old hut. All is OK, and my garden is fine.'

All certainly was 'OK.' That morning, flying the new Nieuport, he had attacked four two-seater reconnaissance aircraft, escorted by a Roland C.II. The enemy formation was flying south-east from St.Léger and Ball picked out the escorting Roland as his first target. He closed the range and fired a full drum into it. Ball's fire was ineffectual. The Roland pilot turned and began to circle the Nieuport, to enable his gunner to bring his rear gun to bear. Ball had changed his Lewis gun drum and, closing to within 20 yards, fired the whole of the fresh drum into the enemy machine. The Roland dived away, the gunner silenced, and force landed in a field on the south-eastern outskirts of St.Léger. Ball turned his attention to the remaining enemy machines, but although he fired two drums of ammunition at them he failed to make any decisive hits and they dived away. Ball was now out of ammunition and he returned.

Ball flew no less than eight missions on 21 August, looking for enemy aircraft without success, but the next day his luck changed. He flew patrols in the morning and afternoon, without sighting any targets, but during the evening patrol, escorting the FEs of the Squadron, he sighted a formation of seven Rolands manoeuvring into position to attack the British two-seaters. Ball flew east of the German machines, cutting off their escape and attacked the rearmost Roland. Ball's fire scattered the rest of the enemy formation, but his target flew straight on. Ball closed the range to 15 yards and opened fire again. The enemy gunner now returned Ball's fire and Ball turned away to change his drum. The fresh drum in position, Ball attacked again, firing half a drum into the right hand side of the enemy machine. The Roland went down vertically and crashed to the west of Bapaume.

Ball then climbed back to the FEs in time to see another five Rolands coming from the direction of Vaux. Ball decided to use his new method of attack. He got under one of the Rolands, pulled the Lewis down its

Ball in SE5 A.4850 at London Colney, March 1917.

slide, and fired a whole drum of Lewis into the forward fuselage of the enemy, concentrating his fire around the pilot's cockpit and the engine nacelle. The Roland flew on. Ball, still in position underneath, changed his drum and fired the whole of its contents into the underside of the Roland. The enemy machine turned slowly, fell away on one wing and went down in flames to crash near Vaux.

Three other Rolands now attacked Ball, anxious to avenge their two comrades, circling the Nieuport and bringing their front and rear guns to bear. Ball turned towards the nearest and fired another full drum into it from a range of 20 feet. Ball's fire wounded both the pilot and the observer of the Roland – *Offizierstellvertreter* Wilhelm Cymera and *Leutnant* Hans Becker of *Kampstaffel* 1. The Roland fell out of control for 6,000 feet and finally crashed into a house near Maurepas. Cymera recovered from his wounds but Becker died.

Ball was now out of ammunition and he evaded the remaining Rolands and returned, landing at La Bellevue to both rearm and refuel. Setting off again to rejoin the FEs, Ball fought a trio of Rolands over Vaux. The Nieuport was hit several times by the enemy's fire, but eventually the Rolands broke off the combat and flew east. The windscreen of the Nieuport had been hit in four places, the rear view mirror broken, and the left-hand spar of one wing hit. His petrol also exhausted, Ball made a good landing only a mile behind the front-line trenches, at Senlis, and spent the night with his aeroplane, flying back the repaired Nieuport the next morning.

Ball was awarded three confirmed victories from this fight. They were the last he was to score in 11 Squadron.

Number 60 Squadron RFC had arrived in France on 28 May 1916, equipped with a collection of Morane types. During the first month of the Somme battle, the Squadron suffered such heavy casualties that General Trenchard had been forced to withdraw it from active duty. On 23 August, having exchanged the majority of their Moranes for Nieuports, the Squadron was relocated at Izel le Hameau, coming under the orders of 3rd Brigade. One of the replacement pilots, posted from 11 Squadron to 60 Squadron was Albert Ball – another was Leslie Foot – and he reported for duty with 60 Squadron on 23 August, taking his Nieuport with him.

Ball soon settled in at his new squadron. Two days after his arrival he shot down a Roland out of control south of Arras and on 28 August he was awarded three more victories. The first was just after 7:00a.m., one of a pair of Rolands from *Flieger Abteilung* 207 flying south-east of Bapaume. Unseen by the enemy crews, Ball took up his favourite position underneath the nearest and fired a complete drum of Lewis into the underside of its fuselage. Changing his drum, Ball fired a further half drum into the Roland from 20 yards range, killing the pilot, *Ltn.* Joachim von Arnim. Arnim's observer, *Ltn.* Böhne, attempted to control the Roland's dive, and finally managed to land the damaged machine near Transloy.

That same evening, flying as an escort for bombers of 3rd Brigade, Ball forced two enemy machines to land, again attacking from underneath in each case. Climbing back from these indecisive engagements, Ball saw two Rolands and two LVGs flying in a loose formation just to the south-east of Adinfer. Determined to score this time, Ball dived at the formation, scattering the enemy machines. He then quickly dived under the nearest Roland and emptied a drum into it from close range. The Roland went down and crashed east of Ayette.

Ball was now the leading British pilot in terms of

victories. Major Smith-Barry, commanding 60 Squadron, asked Ball for a list of his claims. Ball submitted it: 84 combats; 11 machines brought down and one balloon brought down and seen to crash; five machines brought down but not seen to crash; 12 machines forced down and damaged. Ball wrote home: 'So it is not so bad, and I have done my best.' In a letter to his parents at this time, Ball wrote: 'They make it out to be 16 crashed in 84 fights, also the balloon, but I think it is only 12 crashed and the balloon out of 84 fights.'

Ball continued to add to his score and by the end of August it stood at 17. Returning from a patrol on 1 September Ball was told that he had been awarded the DSO. Later in the day 60 Squadron moved its base to Savy, Ball's old aerodrome, reuniting him with his hut and garden. The following day Ball left on leave.

On his return to Nottingham, Ball found that he was a hero, the local newspapers having written at length of his exploits and his gallantry awards. The national press had also taken notice of his success; the *Star* and the *Daily Mirror* reported his rise to fame and the *Daily Mail* printed his photograph, between those of Nungesser, the French ace, and Boelcke, the leader of *Jasta* 2. The *Daily Mail* also ran an editorial on Ball's record, arguing against the official policy of the anonymity of British airmen and suggesting that publicity should be given to the most successful pilots.

In his native Nottingham, Ball was fêted wherever he went, and while he no doubt enjoyed his new-found fame he was embarrassed by the praise given him by men many years his senior.

Returning from leave on 11 September, Ball was given command of 'A' Flight. This was a role which sat uneasily on his shoulders. Ball was not a leader of a Flight; he was not the careful tactician of those fighter pilots who came slightly later – men such as McCudden and Mannock. Ball was essentially a lone fighter, following his instincts, with little thought for tactics or caution.

On 13 September came the notification of a Bar to his DSO and two days later Ball was officially notified that he had been awarded the Russian Order of St.George, gazetted on 24 September.

Ball was now at the height of his fame, firmly acknowledged as the foremost fighter pilot in the Royal Flying Corps. But the air war was changing: the German Air Force was reorganizing, forming its own specialist fighter squadrons – the *Jagdstaffeln*. These *Jasta* were equipped with the new Albatros and Halberstadt biplane D-types, armed with two machine guns firing through the propeller arc. These aeroplanes, although not as manoeuvrable as the Nieuport 16 and 17s, were strong, faster, and, more importantly, carried twice the armament of the British fighters.

Nevertheless, the days of the lone hunter were not quite over. During the next few weeks, until he was posted to Home Establishment on 4 October 1916, Ball added another 14 victories to his score: nine destroyed, including three in flames; two out of control, and three forced to land. This brought his overall victory total to 31, but more important was the example he had set. Ball was Trenchard's offensive policy personified.

After home leave. Ball was posted to 34 Reserve Squadron, based at Orfordness, as a fighting instructor. The order of the War Office was that he was to do no fighting whatsoever, 'even if a Zeppelin passed over his head.' After a few weeks at Orfordness, Ball was posted again, to the School of Aerial Gunnery at Hythe, where he was 'taught' the elementary workings of the Lewis and Vickers machine guns. Ball had joined the RFC to fight and these ludicrously inept postings to a person of Ball's temperament could only have one outcome: he began to pester the authorities to send him out to France again. Like many other pilots, before and after, he also missed the comradeship of squadron life on active duty.

Ball's pleas through the normal channels fell on deaf ears and his request was refused; his present work, he was informed, was the best use to which his experience could be put. Ball then began to pull unofficial strings, using influential friends of his father who he thought could possibly help through their contacts at the War Office. Ball personally wrote to Brigadier-General Charlton, the Director of Air Organisation, but Charlton, although sympathetic, still took the official line. Ball, still not satisfied, carried on his persistent lobbying.

On 18 November 1916, Ball, accompanied by his parents and his sister and brother, attended an investiture at Buckingham Palace and was presented by the King with the DSO and two Bars and the Military Cross. A week later a third Bar to Ball's DSO was gazetted, making him the first officer in the British Army to be awarded three Distinguished Service Orders.

On 23 November, Ball was sent to the Royal Aircraft Factory at Farnborough to test fly a new fighter which had been designed and manufactured at the factory. Ball had no great faith in the Factory products and his first view of the new fighter, the SE5, must have confirmed his prejudice. It looked not unlike the unsuccessful BE12 and a short test flight confirmed his view. He was not favourably impressed and he compared the SE5 unfavourably with the Nieuport, especially in respect of its lack of manoeuvrability in comparison with the little French scout.

On 19 February 1917, Ball was made an honorary freeman of the City of Nottingham for his great services in the Flying Corps. His undoubted embarrassment at the official celebration dinner and the presentation of a silver model of his Nieuport was no doubt tempered by the realization that he was soon to return to France. He had been told that he was to join 56 Squadron, then working up to readiness at London Colney aerodrome in Hertfordshire. What had caused higher authority to at last give in to Ball's requests is not known, but he reported to Major R.G. Blomfield, commanding 56 Squadron, on 26 February 1917.

Blomfield was delighted with Ball's posting. He was determined to make 56 Squadron the best fighter squadron in the RFC and the acquisition of Ball as one of his Flight Commanders was sufficient to give 56 Squadron a reputation before it had even flown in action. This, and the fact that it was to be equipped with the new fighter, the SE5, of which much was expected, made it the talking point of the entire RFC. Blomfield

The first cross, erected by the Germans, on Ball's grave at Annoeullin.

Ball's grave today.

sent Ball on a tour of the flying schools, looking for possible talent, and Ball handpicked several pilots who would later more than fulfill his expectations of them.

56 Squadron left London Colney for France on 7 April 1917, flying to its base at Vert Galant. On the evening of 8 April, Ball flew to Le Hameau to see General Trenchard. Ball had made no secret of his dislike of the SE5 and he begged Trenchard for the use of a Nieuport 17. Trenchard was not unsympathetic and promised to consider the request. On 13 April, Major Blomfield and Ball visited Trenchard, who told Ball that he had authorized that a Nieuport be issued to the Squadron for Ball's personal use, keeping the SE5 for his normal squadron work.

The squadron flew no patrols until the 22 April, the SE5s needing modifications before they could be flown in combat. In the morning, 'A' Flight, led by Ball, left the ground at 10:18a.m. to patrol a line Aix-Neuville-St.Léger. Ball found an Albatros two-seater over Adinfer and attacked it, firing three drums of Lewis, but the enemy pilot dived steeply for his own lines and the SE5s, mindful of their strict orders that they were not to cross the lines, were forced to let it go.

Ball scored his first success with 56 Squadron the next day. He left the ground at 6:00a.m. flying his Nieuport 17, B1522, and made for his patrol area of Douai – Cambrai, knowing the area was thick with enemy aerodromes. Forty-five minutes after takeoff he saw a pair of Albatros two-seaters and he dived underneath them, into his favourite position for attack. One Albatros pilot was too quick for Ball and dived east, but Ball got under the second and fired half a drum of Lewis into it. The Albatros went down and crashed by the side of the Tilloy to Albancourt road. Nobody got out. This was 56 Squadron's first victory and Ball's 32nd. Anxious to score again, Ball attacked another Albatros a few minutes later. The enemy pilot was very experienced and as Ball got underneath him he throttled back and Ball overshot. For a brief moment the Nieuport came into the sights of the enemy pilot and he put a well aimed burst of 15 rounds into two wing spars. Ball dived away and the Albatros flew east.

Ball took off again at 10:45a.m., this time flying the unfavoured SE5. He had an indecisive combat with another Albatros two-seater, a Vickers gun jam forcing him to break off the action. Ball landed at Le Hameau to rectify the jam and took off again, flying to the Cambrai area. At 11:45a.m. he spotted five Albatros single-seater scouts and engaged them over Selvigny, shooting one down in flames.

The remaining enemy scouts then attacked the SE5, putting bullets into the right spar, wings, and two into the fuselage, just behind Ball's head. Ball used the superior speed of the SE5 to shake them off – a tactic he certainly could not have used if he had been flying the Nieuport. Three quarters of an hour later, Ball saw an all white Albatros C.III to the north of Cambrai. He secured his favourite position of underneath, unseen by the enemy crew, and fired half a drum of Lewis into the Albatros. The enemy pilot, *Vzfw.* Ebert of *Flieger Abteilung* 7, dived steeply away from the SE and landed, Ebert helping out his observer, *Ltn.* Berger, who had been severely wounded in the neck by Ball's fire.

The weather on 25 April was bad, and no patrols were flown by the Squadron until the evening of 26 April. Ball took off at 6:15p.m., leading three SE5s. Ball left the other SEs and made for his favoured hunting ground to the west of Douai. Soon after arriving over Cambrai, Ball was fighting for his life. He had attacked a number of Albatros scouts, shooting one down to crash in a wood to the north-east of the town, but was cut off by the remaining five. Ball flew straight at them, but the enemy pilots refused to scatter, boxing him in. Turning continuously, Ball finally managed to break clear and fled to the south-west, pursued by the enemy scouts. One of the enemy pilots, *Vzfw.* Eisenhuth of *Jasta* 3, outdistanced his companions and Ball turned to face him, firing both guns. Eisenhuth's Albatros burst into flames and went down. The remaining German scouts now arrived at the scene but Ball evaded them and returned to base.

Ball next scored on 28 April, shooting down a two-seater from a formation of three, to crash near Fontaine. Ball was separated from the others SEs during this action and he flew to Epéhy. A two-seater came up through the cloud cover and Ball attacked, chasing the enemy machine down through the clouds. Ball was so intent on the chase that he failed to realize how near the ground he was and followed the enemy down to 500 feet. The German anti-aircraft gunners put up a tremendous barrage to aid their countrymen, hitting the SE5 in the engine and shooting some of its controls away. The SE5 went into a spin, but Ball managed to regain control and flew gingerly back to Vert Galant with only his left elevator fully working and that only held by one top wire.

On landing at Vert Galant, Ball was furious. He called for a rag and, wiping the oil from his face, he ordered his Nieuport to be wheeled out. Twenty minutes later the Nieuport left the ground, a still furious Ball at the controls. He returned an hour and a half later, having seen nothing, perhaps fortunately for himself.

Ball next scored on 1 May, shooting down a pair of two-seaters in the last patrol of the day. The following day, Ball took off in the evening, flying with Lt. Knaggs. Between Douai and the front lines Ball saw four red Albatros D.IIIs. He attacked the nearest, but was jumped from the rear by another four. Turning quickly, Ball caused the first Albatros to overshoot and he put 50 rounds into it as it went by. Fastening onto its tail, Ball followed the Albatros down, firing all the time, until it crashed between Halte and Vitry. Ball then flew south to Cambrai and shot down an Albatros C.III. The light was now going fast and he returned.

Ball had his next combat on 4 May, attacking eight Albatros scouts over Riencourt at 12,000 feet and shooting one down to crash near Graincourt.

Ball had a narrow escape the next day. Taking off in the evening, he crossed the front lines and flew north-east at 8,000 feet. Seeing a pair of Albatros scouts flying a thousand feet above him, Ball climbed to 11,000 feet and attacked one of them, shooting it down out of control. The other was now in range and began to attack the SE5. After a brief spell of manoeuvring for a good tactical position, Albatros and SE5 turned to face each other and flew nose on, firing as the distance closed. Both broke away at the last moment, but the SE5 was hit in the engine, covering Ball in oil. Ball zoomed away to review the situation. His engine was still running and although it had lost all oil pressure it was holding its revs. Ball could see no sign of the Albatros and he dived to 3,000 feet. Both the Albatros scouts had crashed and were completely wrecked, lying within 400 yards of each other. Although the Hispano-Suiza engine of the SE5 was getting very little oil, it ran for another three quarters of an hour until Ball finally landed at Vert Galant.

Ball was badly shaken by this combat, telling T.B. Marson, the Squadron Recording Officer, that he was convinced that the Albatros pilot had meant to ram him, and that he had only been saved by divine intervention. 'God is very good to me,' he is reported to have said. 'He must have me in his keeping.'

There can be little doubt that Ball was now dangerously over confident. It also seems that he had not realized the extent to which the air war had changed since he was last at the front in 1916, if indeed, he recognized that it had changed. After Ball's death, 'Billy' Crowe, one of his fellow Flight Commanders in 56 Squadron, commented: 'It was inevitable. The way he fought during that last month made it only a matter of time.'

Sunday 6 May, was fine and Ball scored again, attacking four red Albatros scouts over Douai, sending one down to crash just to the south of Sancourt. It was to be his last victory.

The day started fine and warm on 7 May, but there was a hint of thunder in the air. At mid-day, Ball led his Flight in an escort to the Sopwith 1½ Strutters of 70 Squadron, driving off an attacker on the return flight, but scoring no victories.

The weather deteriorated in the late afternoon, with storm clouds gathering. It was assumed that the worsening conditions would wash out flying for the remainder of the day, but at teatime orders came through from Wing for the Squadron to fly the scheduled evening patrol. At 5:30p.m. 11 SE5s took off and made for their patrol area.

Conditions were now extremely bad, with thick layers of cumulus clouds from 2,000 to 10,000 feet. The SE5s were soon in action with groups of Albatros scouts and the fighting ranged over a wide area of the evening sky, the Flights becoming split up and separated.

Crowe had been in combat with several enemy machines, the last one shooting off his goggles. He came under attack from others and he broke off the action and flew towards Fresnoy. A little later he saw Ball, flying towards Lens. Crowe turned and followed Ball, who fired two red Very lights, the signal that he had seen enemy aircraft. Crowe could see no other aeroplanes in the darkening sky but followed Ball as he flew towards Loos. Suddenly, Ball dived on an Albatros D.III, almost certainly flown by Lothar von Richthofen, getting in a good burst before overshooting. Crowe also attacked and as he turned for another pass he noticed a Sopwith Triplane from one of the Naval squadrons, and a SPAD, both some 500 feet above the action. Turning his attention back to the Albatros, Crowe saw it was again being attacked by Ball. As he watched, both aircraft, still fighting, flew into a thick cloud, and disappeared from his view. It was the last Crowe saw of Ball. He was running out of fuel and he turned away and landed, on his last dregs of petrol, at Naval 8's aerodrome at Auchel.

The next people to see Ball's SE5 were three German officers: a *Leutnant* Hailer and two companions. They saw Ball's SE5 emerge from the low cloud. It was inverted, the engine was stopped and the propeller was stationary. Still inverted, the SE5 crashed in a nearby field. Hailer and his fellow officers hurried to the scene. When they arrived they found that a young French girl had removed Ball's body and it was lying by the side of his wrecked aeroplane. Hailer examined Ball. He could see no bullet wounds and came to the conclusion that Ball had died as a result of injuries sustained in the crash. Hailer and the others then turned their attentions to the wreckage of the SE5. They could see no evidence of damage caused by either aerial combat or anti-aircraft fire. 'We all came to the conclusion that the aircraft had not been shot down in an air fight or by aircraft or anti-aircraft fire as the dead pilot had no marks or scratches and had not been wounded.' As an officer in a *flieger abteilung,* Hailer was well qualified to arrive at these conclusions.

Ball's body was taken to a field hospital, where the examining doctor concluded that he had died of a broken back, severe crushing of the chest, a broken arm, and a fracture of the left leg and foot, all injuries having been caused by the crash.

The following day, *Jasta* 11 telephoned Hailer seeking confirmation of Lothar von Richthofen's claim of a Triplane, crashed in the area. Hailer repeated: 'I did not consider that this aircraft (Ball's SE5) had been brought down, either in aerial combat or by anti-aircraft fire.' In the 1960s, Hailer further stated that revolver bullets were later fired into the wreckage of the SE5, presumably in an attempt to simulate combat damage and give credence to Lothar von Richthofen's claim.

The logical explanation of Ball's death is simply that he became disorientated in the thick cloud, and the SE5 became inverted. The Engineering Officer of 56 Squadron, H.N. Charles, has stated emphatically that the early SE5s could not be flown in an inverted position: the large float-chambered carburetor of the Hispano-Suiza engine would immediately flood the air intake, choking and stopping the engine. This was the position in which Ball found himself when he emerged from the low cloud: the SE5 was inverted, with a dead engine and he had only 200 feet of height. A crash was inevitable.

The Ball memorial on the grounds of Nottingham Castle.

On 8 June 1917 the London Gazette carried the announcement of the award of a posthumous Victoria Cross to Albert Ball; on 21 July, Ball's parents received their son's Cross from King George V at Buckingham Palace.

Ball was buried among his former enemies, in the German military cemetery at Annoeullin. He lies there to this day.

Major William George Barker

Air Chief Marshal Sir Philip Joubert de la Ferté, who as a lieutenant-colonel commanded 14th Wing in Italy in 1918, said of William Barker: 'Of all the flyers of two world wars, none was greater than Billy Barker.' Commenting on this in his book *Canadian Airmen and the First World War,* historian S.F. Wise wrote: 'Despite such praise, and his position as fourth-ranking Canadian ace, Barker's reputation has been overshadowed by others, probably because he won most of his laurels in Italy.'

Barker was born on 3 November 1894 in Dauphin, Manitoba, Canada. At the beginning of WW1 Barker was living in Winnipeg and enlisted in the 1st Canadian Mounted Rifles on 1 December 1914. In June 1915, he sailed for England and was stationed at the Canadian camp at Shorncliffe in Kent. At Shorncliffe, Barker completed a machine gun course and, his regiment now having relinquished its horses, went into the trenches in France as a machine gunner.

As did so many of his fellow countrymen in the Canadian Expeditionary Force, Barker applied for a transfer to the Royal Flying Corps, and in March 1916 he was sent, on probation, to 9 Squadron as an observer under training. Equipped with the BE2c, 9 Squadron had arrived in France on 24 December 1915 and was stationed at Bertangles. On 6 March, Captain H.E. Van Goethem took off in a BE2c on a 20 minute test flight. In the front cockpit was Trooper W.G. Barker.

After several flights for gunnery practice, Barker flew his first patrol over the lines on 12 March, again with Van Goethem as pilot. During a reconnaissance flight lasting an hour and 45 minutes, they took 18 photographs and the results were noted as being highly successful. Two enemy aeroplanes were seen during the patrol but they did not attack.

Barker flew nine patrols with 9 Squadron, his last on 1 April. He was then struck off the strength of the CEF, commissioned as a 2nd Lieutenant in the Royal Flying Corps, and posted to 4 Squadron, based at Baizieux, and also flying BE2cs. Barker flew for three months with 4 Squadron before he was again posted, this time to 15 Squadron, on 7 July 1916. Stationed at Marieux, 15 Squadron was another Corps squadron, also equipped with BE2cs, and it was in this squadron that Barker had his first aerial combats.

With the opening of the Battle of the Somme on 1 July 1916, the corps squadrons of the RFC met fierce opposition from enemy aircraft as they carried out their daily duties of reconnaissance and spotting for the artillery; survival depended largely on their ability to defend themselves by driving off the attacks of enemy fighters. On 20 July, Barker's BE2c was attacked by an enemy scout over Miraumont, but Barker, an excellent shot, drove it off. It was the first of several combats in which Barker fought off enemy fighters.

Even at this stage of his flying career, Barker's offensive spirit was beginning to show itself. In the 1930s an ex-groundcrew member of 15 Squadron, R.H. Arnett, wrote a revealing account of Barker's time in 15 Squadron, both as an observer and later, as a pilot. 'I remember,' Arnett wrote 'that it was not long before he was egging on his pilot to such deeds of "daftness" as might be accomplished in our 2cs and 2es.'

On 16 November 1916, Barker and his pilot, Captain Pender, machine-gunned large numbers of troops in trenches and reported their position to the artillery by wireless. As a result of this and other actions, Barker was recommended for a decoration and a Military Cross was awarded, gazetted on 10 January 1917. 'For conspicuous gallantry in action. He flew at a height of 500 feet over the enemy's lines, and brought back most valuable information. On another occasion, after driving off two hostile machines, he carried out an excellent photographic reconnaissance.'

In late November, Barker was recommended for pilot training and sent back to England. He returned to France – and 15 Squadron – on 23 February 1917 and was allocated to 'C' Flight. Like his fellow countryman Bishop, it seems that Barker was not a good pilot. Arnett again: 'We still had BE2cs and 'C' Flight did not thank him for wiping the undercarriages off three of their machines on his first day! I remember that from that time until we changed over to RE8s, he always flew with a spare undercarriage V strapped on each side of the fuselage and tried hard to carry a spare prop underneath too; but it wasn't feasible, and as we pointed out, it wouldn't have been much good after the old bus had sat on it.' Arnett recalled that Barker specialized in trench strafing and contact patrols and got into and out of many tight corners. '...later, when we got the RE8s, he was brought down again in No-Man's-Land and pancaked quite safely. He had just stepped out of his seat when a shell came right between the V of the engine cylinders and through the machine from end to end, killing the observer on the way and burst just over the tail, without scratching Barker, however.' (No trace of this event can be found in the casualty reports. Author)

Over the next months Barker was mentioned several times in 'Comic Cuts,' the slightly derogatory nickname given to the official RFC communiqués by the members of the corps. On 25 April Barker and his observer, 2Lt. Goodfellow, reported by zone call the presence of 1,000 enemy infantry in trenches, bringing artillery fire down on the enemy troops and two hostile gun batteries. They also took valuable photographs and obtained vital information, all this work being carried out from a height of 500 feet under intense ground fire from rifles and machine guns. Working with the 93rd Seige Battery (5th Army) on 1 May, Barker and Goodfellow destroyed two machine gun nests with five direct hits; working with the same battery, they repeated this the next day, destroying another three nests. Barker and Goodfellow successfully repeated this type of mission throughout May.

On 9 May, Barker was given command of a Flight.

Flight Commanders were normally Captains in rank, but 2nd Lieutenant Barker took charge of 'C' Flight. Two days later he was awarded a Bar to his MC, gazetted on 18 July, for his work during April.

On 7 August Barker was wounded in the head by a shell splinter, which narrowly missed his right eye, but he continued flying until the middle of the month, when he was posted to Home Establishment as an instructor.

Like so many successful pilots, Barker found instructing frustrating and he repeatedly applied for a posting to a fighter squadron, backing up his requests by constantly breaking the orders against low flying. His persistence was finally rewarded and he was posted as 'C' Flight Commander to 28 Squadron, working up to strength at Yatesbury and equipped with Sopwith Camels. While instructing, Barker had flown a Sopwith Camel and had been impressed by its manoeuvrability and speed in contrast to the RE8s he had flown in 15 Squadron. On joining 28 Squadron he was allocated Camel B.6313. This Camel had been taken on 28 Squadron strength on 24 September and Barker first flew it on the evening of 30 September. From this date on, Barker and Camel B.6313 were inseparable, forming a partnership unique in the annals of airfighting.

On 8 October 1917, 28 Squadron flew to St.Omer, and Barker flew a patrol over the lines that evening, his first as a fighter pilot. Some reports give that Barker shot down an Albatros during this patrol, but if so he did not make a claim and it was not until 20 October, with 28 Squadron now stationed at Droglandt, that he claimed his first victory. A combined force of Camels from 70 and 28 Squadrons attacked the enemy aerodrome at Rumbeke, the base of *Jasta* Boelcke, and Barker shot down an Albatros D.III in the fighting.

Four days later, Barker attacked a formation of 'Gotha' bombers over Ypres. From 200 yards he fired 150 rounds into the rearmost machine, halving the range before firing another 100 rounds. Barker's guns then jammed and he broke off the combat, the enemy machine gliding down from 7,000 feet with its starboard engine stopped.

On 26 October, although the weather was very bad, with wind and rain, Barker took off with Lts. Mailik and Fenton. The Camels were soon in combat with four Albatroses from *Jasta* 18, flown by *Ltn.* Schober, *Ltn.* Rahn, *Ltn. d.R.* Paul Stähle , and *Offizierstellvertreter* Klien. Barker and Stähle fought for 15 minutes and Stähle later wrote in his diary, 'In a tough dogfight lasting over a quarter of an hour, sometimes only a few feet above the ground, I fought the enemy scout as far as Ichteghem where unfortunately I had to break off because my guns had jammed. For me this fight was the hottest and most exciting that I had in my whole fighting career.'[1] From Stähle's account of the fight it is obvious that he was the enemy pilot who fought Barker for 15 minutes, but although he landed safely on Ichteghem aerodrome the RFC communiqué states that the Albatros from this long fight went down in flames and was seen to crash, presumably quoting from Barker's combat report. Stähle's diary only reports one loss to *Jasta* 18 on 26 October, *Ltn.* Schober, and although he records that Schober was attacked and shot down by 'the other Englishman,' Barker was credited with two victories from the fighting. These two victories were the last Barker was to score in France in 1917.

1. *Cross and Cockade International,* Vol.11, No.4.

Captain William George Barker while serving with 15 Squadron. The aeroplane is an RE8.

On 24 October 1917, Austro-Hungarian and German forces attacked the Italian positions at Caporetto. The Italian troops, demoralized by 11 offensives on the Isonzo River between June 1915 and September 1917, each of which had caused them heavy casualties with little success, fell back over 70 miles until they managed to hold the enemy's advance at the Piave River. The Italian armies suffered over 200,000 casualties, with another 400,000 troops deserting, and badly needed help from their allies. The British sent five divisions to Italy, the French another six. Included in the British forces were 28 and 34 Squadrons of the RFC, and a new Wing, the 51st, was formed under the command of Lt.Col. R.P. Mills, DSO.

On 29 October, 28 Squadron moved to Candas in preparation for its move to Italy, leaving France, by train, on 9 November. While 28 and 34 Squadron were in transit – 34 Squadron had left on 7 November – it was decided to reinforce them with an additional three squadrons and 45, 66, and 42 Squadrons were selected. 42 Squadron left Candas on 26 November, but 45 and 66 Squadrons, delayed by the beginning of the Battle of Cambrai, left on 11 and 12 December respectively. On 18 November, V11 Brigade was formed, comprising the 14th and 51st Wings, under the command of Brigadier-General T.I. Webb-Bowen, and 28 Squadron, along with 45 and 66 Squadrons, came under the orders of 14th Wing.

In contrast to their comrades on the ground, the fighter pilots of the Italian Air Force were highly motivated and well trained, flying SPADs, Nieuports, and the newly introduced Hanriot fighters. However, as part of the reinforcements to their Austrian allies during the Caporetto offensive, Germany had transferred three *Jadgstaffeln* from the Western Front, plus seven *Flieger Abteilung*. These units were highly experienced and soon made their presence felt.

Arriving at Milan on 12 November, 28 Squadron unpacked its crated Camels and had them assembled and tested within two days. After moving to Ghedi on 17 November, then to Verona on 22 November, the Squadron finally arrived at the front on 28 November and were based at Grossa.

The first patrol over the lines by the squadrons of the RFC took place on the late morning of 29 November 1917, when an RE8 of 34 Squadron, escorted by three Camels from 28 Squadron, flew a photo-reconnaissance in the area of Montello. Five Albatroses attacked the British formation at 10,000 feet before it had even crossed the front line, and during a fight lasting some 20 minutes, the enemy fighters were reinforced by seven more. Barker shot the wings off an Albatros which crashed in the area of Pieve di Soligo.

Barker reported that during the combat, at 10,000 feet, his Camel had been outclassed by the Albatros in respect of both speed and climb, a situation which was only reversed when the fighting dropped to 5,000 feet. Barker identified these Albatroses as D.IVs, but the Albatros D.IV was an experimental prototype which never went into production, and the Albatroses in the combat were probably Öeffag-built Albatros D.IIIs, These aeroplanes were powered with the 185hp Austro-Daimler engine – more powerful than the 160hp Mercedes of the German built Albatros D.III – which would explain their superior performance to Barker's Camel.

On 3 December, Barker shot down a balloon and an Albatros. After escorting the RE8s of 34 Squadron back over the Lines, Barker, with two other Camels flown by Lts. Cooper and Woltho, crossed the River Piave at low altitude and attacked a hostile balloon. Barker attacked from 1,000 feet and fired 40 rounds into the balloon, which was being winched down. Turning away from the balloon, Barker then saw Stanton Woltho about to be attacked by an Albatros and dived to his aid. He drove the enemy scout down to 300 feet and fired a long burst. The Albatros went down vertically and crashed, the wreckage bursting into flames. Barker again attacked the balloon, by this time on the ground, and set fire to it. After firing on the winch crew, Barker attacked a large staff car coming east from Conegliano, forcing the driver off the road, overturning the car into a ditch.

On Christmas Day, Barker and a fellow Canadian, Lt. H.B. Hudson, attacked the enemy aerodrome at Motta, ten miles behind the enemy lines, and the home of *Flieger Abteilung (A)* 204. Barker and Hudson shot up the hangars and damaged four aircraft. As a passing shot they dropped a large piece of cardboard inscribed 'To the Austrian Flying Corps from the English RFC. Wishing you a merry Christmas.' This attack brought a retaliatory raid on 34 Squadron's aerodrome at Istrana on 26 December, Boxing Day. A member of 34 Squadron later wrote.

'After his famous raid on Christmas Day, Barker landed on our aerodrome before going on to his own and it was thought that this was observed by, or conveyed to, the Austrians, hence the retaliation raid the following morning. This took place at about 9:00a.m., not 7:00a.m. It was not a very good morning for observation, and with the exception of two machines out on the aerodrome, all our RE8s were in their portable canvas hangars. The first intimation of the impending raid was the opening up of a nearby Archie battery which up to then had not been heard. The raiders were then a short distance from the aerodrome swiftly approaching it in a dive.

We had not any trenches, dugouts, or sandbag "dives" and the workshop lorries were parked end to end in a line directly across the line of approach of the Huns (sic). Altogether a most attractive target. The hangars were more or less in line with the line of approach, but the Italians were disposed in a semi-circle more or less across the approach.

We were thoroughly shot up and bombed and not only were two RE8s wiped out, but all the squadrons were unserviceable for hours, being more or less riddled. Inside the hangars gave one the idea of a starlight night, so holed were they.

Our orderly room sergeant was killed outright by a flying bomb fragment and we also had seven wounded. The spectacle of our fellows being chased by machine gunners in planes and the observers aiming hand bombs at us was not good.

I have the greatest admiration for the Italians who got two scouts off in the midst of the raid...Their hangars were badly damaged and at least one fired.'

Between 30 to 40 aircraft took part in this raid – some eyewitnesses counted 42 – and they attacked from heights ranging from 500 to 3,000 feet, some of them dropping their bombs before reaching Istrana. After the raid the Austrian machines did not fly straight back to their own Lines, but wandered 'all over the country at about 1,000 feet,' and four were shot down on the allied side by Italian pilots and anti-aircraft fire, with Italian pilots claiming another three brought down on the enemy side. Another, smaller raid, was attempted just after mid-day, but this was turned back.

One of the pilots captured from the first raid stated that the order for the attack had come from their headquarters at breakfast time, when all the aircrew were sleeping off the effects of the Christmas Day celebrations, most only having been in bed for a few hours. One aeroplane had not even been refueled before starting on the raid and ran out of petrol a few miles to the west-north-west of Istrana aerodrome. The pilot landed safely and then promptly fell asleep where he sat. He was later woken by the touch of his own pistol – in the hands of an RFC officer.

On 29 December, Barker shot down a balloon near Pieve di Soglio for his seventh victory, and he opened the New Year of 1918 with a victory over an Albatros, on 1 January. The RE8s of 42 Squadron had bombed the

Barker with his favourite Camel B6313, in which he scored 46 of his 50 victories.

Austrian Army Headquarters in the town, escorted by five of 28 Squadron's Camels, led by Barker. Enemy fighters came up to intercept the RE8s and a fight started at 11,000 feet. Barker shot down an Albatros D.III, flown by *Offizierstellvertreter* Karl Lang of *Jasta* 1, the Albatros crashing into the mountainside to the north-west of Vittorio, bursting into flames and rolling down into the valley. Lang was killed.

It was nearly a month before Barker scored again, on 24 January. In company with Lt. Hudson, he shared in the destruction of two balloons shot down near Conegliano. Barker and Hudson had taken off on what were ostensibly only practice fighting and machine-gun tests, but Barker's combat report reads. 'While testing guns over the Lines we sighted two balloons in a field which we attacked and destroyed in flames. A horse transport column of about 25 vehicles which was passing these balloons was also attacked and stampeded.' It was obvious that the Camels had attacked the balloons from low altitude, which was strictly against orders, and Lt.Col. Joubert, now commanding 14th Wing, demanded an explanation – in writing. Barker's excuse was that, on sighting the balloons, he had for the moment forgotten the orders forbidding low flying.

On 2 February, on patrol at 14,000 feet, again in the vicinity of Conegliano, Barker's Flight sighted five enemy two-seaters, escorted by three scouts. The Camels attacked. Lt. C.M. McEwan, from Manitoba, shot down two of the two-seaters, and Barker dispatched another. Barker then attacked one of the escorts, a Phönix D.II, which he send down to crash in a field. Seeing a crowd of enemy troops gathering around the crashed Phönix, Barker dived and set the wreckage on fire, scattering the enemy infantry.

Three days later, Barker and Hudson took off to look for an enemy two-seater, reported to be working over the lines at 17,000 feet north of Odense. They found the two-seater, escorted by two Albatroses. Barker shot the left wing off one Albatros, which broke up in mid-air, then fought the two-seater down until it landed in a field and turned over. Hudson had shot down the other escort, fighting it down to 200 feet until it finally crashed.

Bad weather stopped nearly all flying for the next few days. The pilots of 28 Squadron were soon bored with the inactivity. A sweepstakes was organized. The names of all the pilots were put into a hat. The pilot who drew the name of the next to score would scoop the jackpot. The next day the weather was still very bad, with thick mist, but Barker and Hudson were convinced that conditions were clearer at height and took off. The Camels flew to Fossamerlo where they knew three small spherical balloons were parked in a line near the ground, with a standard type of observation balloon at each end. Barker and Hudson destroyed all five balloons. Arriving back at Grossa, Barker asked if balloons counted in the sweep. On being told they did, he dryly commented that whoever had drawn his name in the sweepstakes should collect his winnings. All five balloons were shared with Steve Hudson.

Barker then left on three weeks' leave and while enjoying a period of well-earned rest the award of a DSO was announced on 18 February 1918.

Barker returned to Italy on 8 March. It was not a happy start. The day after his return he crashed his favourite Camel, B.6313, which he had flown since its delivery to 28 Squadron the previous September. He had continued to fly this Camel in France in October and had scored all his victories, both in France and Italy, while flying it.

Barker had crash landed, due to engine failure, and B.6313 was badly damaged: it needed a new set of wings and struts, center section, rudder bar, a diagonal front strut in the fuselage and the wrecked undercarriage had to be replaced.

Barker was deeply attached to this Camel – despite the fact that many pilots would have been unhappy to fly an aeroplane whose serial number added up to 'unlucky' 13, compounded by the figure 13 in the serial number itself – and he may well have waited for the Camel to be repaired before flying again. Whatever the reason he did not score again until 18 March, ten days after his return. Barker took off after lunch on 18 March and shot down an Albatros over Villanova, shooting down two more the next day: one out of control, the other destroyed. He was flying B6313.

These last two victories brought Barker's total to 22, making him the highest scoring RFC pilot on the Italian Front, and when Major Glanville relinquished command of 28 Squadron on 18 March, Barker, as senior Flight Commander may well have thought that the natural chain of command would pass to him. Instead, the command of 28 Squadron was given to Major C.A. Ridley, DSO, MC, who arrived from England. Pilots in Italy at the time, remember that Barker was disappointed at being passed over and less than happy with the new arrangement. He asked for a transfer to another squadron, and Captain J.M. Whitlock, commanding 'C' Flight in 66 Squadron, exchanged places with Barker on 10 April.

The change of squadron seemed to make no difference to Barker. Still flying his favourite Camel B.6313, which he had taken with him to 66 Squadron, he opened his score with the Squadron on 17 April, shooting down an Albatros east of Vittorio. But the weather for the rest of April was bad, with few patrols being flown, and it was not until nearly a month later, on 8 May, that Barker scored again, a two-seater destroyed over Annone-Cessalto. By the evening of 23 May, Barker had scored another five victories, bringing his total to 29.

On 24 May, *Offizierstellvertreter* Jozsef Kiss, a leading Austrian ace with 19 victories, accompanied by *Feldwebel* Alexander Kasza and Kirjik took off from his aerodrome at Pergine, the base of *Flik* 55J. Their orders were to intercept a force of Caproni bombers that were heading towards Feltre and Belluno. At approximately 10:35a.m. the Kette saw three Sopwith Camels near Grigno.

That same morning, Barker had taken off with Lts. G.A. Birks and G.F.M. Apps at 10:40a.m. Patrolling over the Eastern area, they sighted two Albatros scouts and one Aviatik (commonly called a Berg fighter) over Grigno at 17,000 feet. Barker attacked the rear machine, which spun away from his fire. Meanwhile, Birks had attacked the Berg and shot its wings off after a short fight. This was seen by Barker, who then saw that three additional Albatroses were diving from the south to attack Birks and Apps, who were fighting with 'the remaining two EA in the valley.' Presumably these 'remaining two EA' were Barker's first opponent, who had climbed back into the fight, and the other Albatros.

Barker got under the tail of one of the arriving Albatroses and shot it down out of control. It crashed onto some hutments in the valley and burst into flames. This was seen and confirmed by Apps and Birks. Birks was now attacking one of the original Albatroses, but its companion from the first formation was on his tail. Apps saw the danger and shot the Albatros off Birks' tail, firing a long burst into it. The Albatros went down out of control to crash in the valley. This was confirmed by both Barker and Birks; Barker seeing it go down out of control and Birks witnessing the crash.

Birks was now free to fight the remaining Albatros from the first formation, but the enemy pilot was 'exceptionally skillful' and Birks fought him for a long time until joined by Apps. The Albatros pilot was still a match for the two Camels. Barker then joined the fight, got on the tail of the Albatros and shot it down out of control to crash into the same hutments as his first victory.

The Camel pilots claimed four enemy machines from this fight: two by Barker and one each to Apps and Birks. *Flik* 55J had only one loss that morning: Kiss, who was shot down and killed. It has not been possible to establish the identity of the unit of the other Albatroses arriving at the scene of the fight, one of which was claimed by Barker, but leaving their part in the fight aside for the moment, the details of the fight, as given in the 66 Squadron combat report, fits the Austrian account of the combat in which Kiss was lost. Kasza was considerably shot about by the Camels, but landed safely at Feltre, which fits with the Camel pilots' report that the only enemy machine seen to escape from the fight was last seen, very low down, in the direction of Feltre. Kirjak, the third member of the *kette,* managed to land back at Pergine, his aeroplane also badly damaged in the fight.

Accepting that Barker's first victory, which was seen to crash into the hutments in the valley, was not from *Flik* 55J, but from the other formation of Albatroses, the remaining three victories claimed by the Camel pilots were the 'Berg' shot down by Birks at the beginning of the fight, the machine shot off Birks' tail by Apps, and the exceptionally skillful pilot later claimed by Barker. From the Austrian account it would appear that this last was Kasza, who was a very accomplished aerobatic pilot. It has been stated that these three pilots of *Flik* 55J operated a combat tactic of both Kiss and Kirjak diving away if attacked, as if running from combat, while Kasza stayed to fight, relying on his superior flying skills to avoid being shot down while his comrades climbed back, unseen, into the fight, to take the attackers from the rear. This scenario fits the 66 Squadron pilots' report of the fight: Barker shot at the first Albatros (Kirjak) which spun and dived away, as to plan. But before the 'Berg' (Kiss) could follow, Birks shot it down. This left Kirjak, climbing back into the fight, and Kasza.

The Camel pilots were mistaken in claiming that both Kasza and Kirjak were seen to crash, but it must be said that their combat reports were not signed by them, but by 66 Squadron's Recording Officer. Unless the last two enemy machines claimed by Barker and Apps were from the other formation of Albatroses, which, from the reports of the fight seems unlikely, then the Camel pilots over-claimed by two.

It has been stated in the past that Kiss, Kirjik, and Kasza were all flying Phönix D.IIs on the morning of 24 May, but this is at variance with the Camel pilots' report

Two views of Barker's famous Camel B6313. The front view shows Barker's tally of victories marked on the left forward wing strut. In the rear view the heart, pierced with an arrow, is clearly visible.

that they engaged two Albatroses and a Berg. There is a photograph, dated 8 May, which shows Kiss with his Albatros D.III (Oef) in *Flik* 55J, but by early June the *Flik* was equipped with the Phönix D.IIa. It is known that Kiss was killed in Phönix D.IIa 422.10 and it is possible that Kiss, in his capacity as a 19-victory ace, was flying one of the first of the new type to be taken on charge by *Flik* 55J, his two companions still flying their Albatros D.IIIs. The Phönix was a new type, not dissimilar to the Aviatik D.I, (known as a 'Berg' by the allied fighter pilots) and the 'Berg' claimed by Birks could well have been Kiss' Phönix D.IIa.

Barker next claimed a victory on 3 June, destroying a Berg C-type over Fiume at 8:30a.m., following this with two victories on 9 June over two Berg fighters within five minutes of each other, the first at 10:20a.m. over Levico. By 13 July, two more Berg fighters and two Albatroses had fallen under his guns, bringing his total victories to 38.

Not all Barker's fighting was in the upper air. On 15 June the Austrians launched an attack along the front from the Adriatic to the Asiago Plateau. The low mist hampered RAF support for the hard pressed ground troops, but 66 Squadron had offensive patrols out from 4:35a.m. Barker, leading Birks and Bell, bombed targets in the area of Val d'Assa, but by 9:00a.m. the mist had thickened and air operations had to cease. Conditions improved later in the day and the RAF squadrons operated over the Piave Front, ground strafing and bombing.

Barker led an attack on two pontoon bridges in the Montello sector. He selected the bridge which was farthest upstream and starting bombing from 50 feet. The bridge was hit in two places, the pontoons broke loose and, caught by the fast current, smashed into the pontoons of the lower bridge, carrying it away. At the time of the attack the bridges were crowded with Austrian infantry and they were thrown into the water. Those that managed to get into small boats or onto islands were machine-gunned by the Camels. Barker and four 66 Squadron Camels destroyed another bridge the next day, but it was reported in the early afternoon that seven bridges on the lower stretch of the river were still intact.

Sometime during June, Barker issued what S.F. Wise, the Canadian air historian, has called an 'absurd and vainglorious challenge.' Barker, Birks, and McEwan dropped a note on Godega aerodrome, challenging Godwin Brumowski, Austria's leading ace; Benno Fiala *Ritter* von Fernbrugg, and Friedrich Navratil to meet the pilots of 66 Squadron for an aerial duel. The challenge was not taken up.

There was a long pause in the war on the Italian Front from late in June until October, but the RAF kept up its offensive policy. On 3 July a new squadron was formed, 139 Squadron, equipped with the Bristol Fighter, and with its personnel drawn from 34 Squadron and drafts from England and France. On 14 July Barker was promoted to Major and given command of 134 Squadron, again taking his favourite Camel B6313 with him to his new command.

Captain Wedgewood Benn (the father of the present [1997] Labour member of parliament, Tony Wedgewood Benn) had been an observer with 34 Squadron, flying in RE8s, but he had left the Squadron to become the 14th Wing Intelligence Officer. The Italians had conceived the idea of dropping a spy by parachute into enemy-held Italy to obtain information from the local people of the area. Benn was enthusiastic about this idea and a Caproni bomber was selected as the type of aeroplane to be used. Tests showed that the Caproni was unsuitable, and a twin-engined Savoia-Pomilio S.P.4 was finally chosen and hangared at 134 Squadron's aerodrome.

Barker and Benn flew a reconnaissance on the night of 28 July, flying a Bristol Fighter over the route of the drop. On the night of 9 August, Barker, Benn, and the spy, Alessandro Tandura, took off, flew to the arranged area, and successfully dropped Tandura. The whole mission was a complete success and Tandura was later awarded the Italian Gold Medal for Valour – the Italian equivalent of the British Victoria Cross – for this and other missions. Barker and Benn were both awarded the Italian Silver Medal for Military Valour on 12 September 1918.

Barker did not let his new duties and responsibilities as a commanding officer keep him on the ground and out of combat. On patrol in B6313 on 18 July, accompanied by a Bristol Fighter, Barker shot down an LVG from a formation of three enemy scouts escorting a pair of two-seaters. Two days later, with two Bristols from his Squadron, Barker saw six Albatros D.IIIs over their aerodrome at Motta. Three Italian scouts then appeared and the Albatroses all dived for the safety of their base. Barker and the Bristols dived with them and were apparently mistaken for friendly machines, as the Albatroses pulled out of their dives and began to form up with Barker's formation. Barker attacked an Albatros, which broke up after 40 rounds. He then fought another down to ground level until it crashed. One of the Bristol crews accounted for another Albatros, confirmed by Barker.

Barker shared in the destruction of another enemy machine on 23 July, shot down by himself and Lts. Walters and Davies in a Bristol Fighter. The same day he was awarded a Bar to his DSO.

Barker scored no victories in August although he flew several Offensive Patrols. On 16 September 1918, Barker had an important passenger in the person of Major His Royal Highness, Edward, Prince of Wales, who flew in the observer's cockpit of Bristol Fighter D7972 for 45 minutes.

On 18 September, Barker gained his last victories in Italy. On Offensive Patrol in the morning he saw nine Albatros scouts at 9:50a.m., flying west at 16,000 feet. Being heavily outnumbered, Barker attempted to get some Italian fighters in the vicinity to attack with him, but only one, a Nieuport with a red fuselage, followed him. Camel and Nieuport flew parallel with the enemy formation until about four miles south of Feltre, when, at 10:20a.m., Barker gave the signal for the Nieuport to attack. Barker attacked one of the Albatroses at 17,000 feet and shot it down out of control opening fire at a range of 60 yards, closing to within ten yards for a final burst. Barker then came under attack from two of the enemy scouts but he half rolled away and shot at another from close range. This went down out of control. Up to this point Barker had noted that the Albatroses were not very well flown but he was then attacked by a very skillful pilot, who flew head on towards the Camel. Barker broke in a climbing turn at 40 yards and got on the tail of the Albatros. No other Albatros interfered in the fight and the Albatros dived away from Barker's fire. Barker followed and got in a long burst from 100 yards range. The Albatros turned over onto its back, began to smoke, and the pilot jumped out and was last seen parachuting down, his Albatros crashing into a river bed three miles north of Quero. Photographs were later taken of the crash to confirm Barker's claim.

On 29 September Barker made his last two flights in his beloved Camel B6313. The Camel had flown a total of 404 hours 10 minutes, of which 326 hours 30 minutes were on operational sorties – a remarkable number of hours for a fighter in WW1 – and Barker had scored all his victories while flying it. Later that evening B6313 was dismantled by 'B' Flight, 139 Squadron and taken complete – except for the clock which Barker had kept for a souvenir – to No.7 Aircraft Park. The following day, Wing sent an order to Barker to return the clock.

Barker left Italy on 30 September, posted to Home Establishment to take command of a school of air fighting. On leaving Italy his victories totaled 46. He had been awarded a DSO, gazetted on 18 February 1918; been Mentioned in Despatches on 8 May; a second Bar to his MC had been gazetted on 16 September 1918, and a Bar to his DSO was gazetted on 2 November 1918.

On arrival in England, Barker requested that he be attached to a squadron on active duty in France before taking up his new post, arguing that tactics in Italy differed from those in France and he would need a refresher course in order to do full justice to his new command. Barker's request was granted and flying one of the new Sopwith Snipes, E.8102, he was attached to 201 Squadron for ten days, arriving on 17 October.

Barker flew three uneventful patrols with 201 Squadron on 21, 22, and 23 October, seeing no action. On 27 October, his tour ended, Barker took off to fly back to England. He could not resist one last look at the fighting. He climbed to 21,000 feet over the Forest of Morval and attacked a Rumpler, which broke up under his fire. Barker then came under attack from below. A Fokker D.VII from a formation below Barker had climbed until directly underneath him, hung on its prop in a near stalling position – a manoeuvre much used by D.VII pilots – and fired into the Snipe, wounding Barker in the right thigh. Barker threw the Snipe into a spin for 2,000 feet, but on pulling out he found himself in the middle of a large formation of Fokker D.VIIs. He fired at two Fokkers, but they evaded him. Barker then fired into another at close range and set it on fire, but he was now attacked by all the remaining Fokkers and wounded again, this time in his left thigh. Barker fainted and the Snipe fell into a spin. Barker recovered consciousness at 15,000 feet, only to find that he was now in the midst of a lower formation of Fokker D.VIIs. He attacked one

Barker's Sopwith Snipe E8012 after his epic fight of 27 October 1918.

Fokker, getting on its tail and shooting it down in flames, but there was another Fokker on the tail of the Snipe and its fire hit Barker in the left elbow, shattering it. Barker again fainted with the pain of his wounds and the Snipe again spun, this time falling another 3,000 feet before Barker came to. He was yet again surrounded by a formation of Fokker D.VIIs. The Snipe was now badly damaged, with its engine smoking, and Barker determined to finish the unequal fight by ramming the nearest Fokker. He flew straight at it, firing as he closed. The Fokker broke up in mid-air and the Snipe was further damaged as Barker flew through the wreckage. Barker was then momentarily in clear air and he dived for the safety of the British front line, flying under another formation of enemy aircraft. Barker crossed the front line trenches at tree-top height and crashed into barbed wire surrounding an observation balloon site.

One of the witnesses – the entire action had taken place in full view of thousands of British and Canadian front-line troops – was A.G.L. McNaughton, commanding the Canadian Corps Heavy Artillery, who watched the fight from his headquarters between Bellevue and Valenciennes. 'The hoarse shout, or rather prolonged roar, which greeted the triumph of the British fighter, and which echoed across the battle front, was never matched...on any other occasion.'

Barker was taken from the wreckage and rushed to hospital in Rouen. Ten days later he was well enough to ruefully admit 'By Jove, I was a foolish boy, but anyhow, I taught them a lesson. The only thing that bucks me up is to look back and see them going down in flames.'

Barker was awarded the Victoria Cross for this last fight, receiving his Cross from King George V on 30 November 1918. The war had been over 19 days.

After the war, Barker returned to Canada. He left the RAF in 1919 and went into civil aviation, forming a partnership with another Canadian VC, William Bishop. After the failure of this enterprise, Barker joined the Canadian Air Force and served from 1920 to 1924. Barker was then involved various business ventures until January 1930, when he became Vice-President of The Fairchild Aviation Corporation of Canada. On 12 March 1930, William George Barker VC, DSO and Bar, MC and 2 Bars, C.C., L.d'h., C.de G., M.d'Arg., was killed in a flying accident while demonstrating a prototype aeroplane at Rockcliffe Airport, Ottawa. Thousands attended the funeral in Toronto.

From his action which won him the Victoria Cross, William Barker is chiefly remembered as a lone wolf type of fighter pilot, 'something of a gambler.' However, from the memories of those who knew and flew with him, a different picture emerges. Both as a flight commander and squadron commander, Barker carefully trained the less experienced pilots under his command, often giving victories to others which he could have claimed for himself. It is true that he excelled in individual fighting, and was one of the few fighter pilots who had no fear of ground strafing, but he was also a highly skilled patrol leader. Above all, and like James McCudden, Barker was a first rate shot. Captain James H. Mitchell DFC, who served in 28 Squadron with Barker in both France and Italy, wrote: 'Whilst one could not say he was a good pilot he certainly made up for this in his shooting...I have seen enemy machines break up in the air or go down in flames long before I realized we were in range. He dispensed with the Aldis sight and put a piece of wire across the muzzles of the gun barrels with a central bead, and a similar one on the rear with a V-sight and these suffice...He was most abstemious and temperate and a non-smoker.' Although Barker was one of the most decorated pilots in the RAF, none of his contemporaries considered that he was in any sense a medal hunter. Barker was the last Canadian airman to win the Victoria Cross in the First World War.

Lieutenant Richard Bell-Davies

Richard Bell-Davies was born on 19 May 1886 and enlisted in the Navy when he was 15 years of age. Inspired by a flight over the fleet by the pioneer aviator Claude Grahame-White, the young officer applied for membership of the Royal Aero Club and took private flying lessons at the Grahame-White school at Hendon, gaining his RAeC certificate No.90, on 19 May 1911. Bell-Davies was then posted to the China station but he had applied for pilot training and was sent back to England to commence his training. Having completed this, he was promoted to the rank Flight Commander in the Naval Wing of the RFC on 10 July 1913 and by the end of the year was a Squadron Commander.

In 1915 Bell-Davies was serving in France in 3 Squadron RNAS under Commander Charles Samson, and on 23 January 1915 he and his observer, flying Maurice Farman No.1241, took off to attack a submarine in Zeebrugge harbour. Bell-Davies crossed the coast at Nieuport and headed inland, but came under heavy anti-aircraft fire. The Farman was hit several times and Bell-Davies was hit in his right leg. He flew out to sea, taking the Farman out of range, then, despite loss of blood and being in extreme pain, he turned back and re-crossed the coast. Arriving over Zeebrugge his 20lb Hale bombs straddled the submarine.

After his wounds had been attended to Bell-Davies was sent back to England. For his attack on the submarine, and his previous work in France, he was awarded the Distinguished Service Order, gazetted on 9 April 1915.

In March/April 1915, 3 Squadron was moved to Tenedos: its task was reconnaissance of the Straits, and on 4 August 1915 the squadron was moved to Imbros to support the landings at Sulva Bay.

By the end of October 1915, 3 Squadron was carrying out long-range bombing raids in attempts to cut the Turkish supply lines from Germany, which passed through Bulgaria, concentrating its attacks on railways and roads. On 5 November, the Admiralty urged the RNAS at Imbros to attack the railway bridge over the River Maritz in Eastern Thrace. If the line could be cut at this point it would interrupt communications from Berlin to both Constantinople and Salonica. The RNAS attacked the target six times during daylight and once at night during the month. The bridge was never hit but an attack by Commander Samson on 8 November damaged one of the bridge pillars and delayed traffic for two days.

On 19 November, Commander Samson decided to carry out a full-sized attack on the railway at two key points: Burgess Bridge and Ferejik Junction. The entire squadron took part, Bell-Davies in Nieuport 10 No.3172, being given Ferejik as his target. With a bomb load of six 20lb bombs and with the front cockpit of the Nieuport covered, Bell-Davies took off for Ferejik, just inside Turkey's border with Bulgaria, near the Maritza River and some 15 miles from the Gulf of Xeros.

Bell-Davies approached the target in company with Lieutenant G.F. Smylie, flying a Henry Farman coded H 5. They found the railway junction and dropped all but one of their bombs, but the Farman was hit by rifle fire and Smylie was forced to land near the bridge. The whole area was marshland, inundated with water channels of dry mud baked hard by the summer heat and the Farman came down in one of these. Smylie ran from his machine and hid from the advancing Bulgarian troops in a small hollow. Bell-Davies saw Smylie's plight and decided to land and pick him up. The Nieuport was capable of carrying two people but the state of the ground would make takeoff difficult, if not impossible, and Bell-Davies was unsure how many crew the Farman had been carrying. Smylie saw that Bell-Davies was coming in to land. Remembering that the Farman still had a bomb on board he ran back to it and exploded the bomb with a shot from his pistol. This alarmed Bell-Davies and he climbed away until he was sure that there were to be no other explosions. Returning, he made a good landing and as the Nieuport rolled to a stop in the mud, Smylie ran from his cover, swung the wing round

Richard Bell-Davies.

This Nieuport 10 is believed to be the machine in which Bell-Davies won his VC. It is seen here in the earlier tail marking used by the RNAS and is coded 26 in the rudder marking. At the time of his VC action, Bell-Davies' Nieuport was marked as shown in the colour plate.

and ran alongside, holding the aeroplane steady while Bell-Davies taxied back between the reeds for a takeoff into wind. Smylie then jumped into Bell-Davies' cockpit. Smylie was six feet four tall but he managed to crouch on all fours on the floor of the cockpit with his head against the petrol tank. With the Bulgarian troops now within range and beginning to fire on the Nieuport, Bell-Davies opened up. The little Nieuport raced along the marshy ground at full throttle and to the relief of both Bell-Davies and Smylie finally became airborne. Bell-Davies set course for Imbros, landing some 45 minutes later.

Commander Samson recommended Bell-Davies for a Victoria Cross for this action and this was gazetted on 31 December 1915. Gilbert Smylie received a Distinguished Service Cross.

Bell-Davies rejoined the Navy after the war and at the outbreak of WW2 he was a Rear Admiral. In May 1941 he was promoted to Vice-Admiral, retiring from the service in 1944. On 26 February 1966, Vice-Admiral Richard Bell-Davies, VC, CB, DSO, AFC died in the Royal Naval Hospital at Gosport.

Lieutenant-Colonel William Avery Bishop

William Avery Bishop was born on 8 February 1894, at Owen Sound, Ontario, Canada, the youngest son of William and Margaret Bishop. As a youth, Bishop had a poor academic record and, knowing this would preclude his entering university, he decided on a military career. Entrance to the Royal Military College was by examination – rather than academic achievement – which Bishop felt he could pass. His elder brother, Worth, had achieved the highest standing of any student in the history of the college and this, together with the fact that he was a first class shot and a good rider, were factors which Bishop felt would sway the college authorities to accept him. His school principal, however, was not encouraging. When told of Bishop's plan he dismissed the idea, bluntly telling the young Bishop that he did not have the intelligence to pass the examination, but after a great deal of cramming, Bishop sat the entrance exam in 1911 and won a place at the College.

Bishop failed his First Year examinations, but was allowed a second chance. He worked hard for a year and passed at his next attempt. His third year at the college, however, was a disaster. Bishop was a natural rebel and a second year of enforced good behaviour was too much: his third year at the college was a litany of broken rules, disciplinary crimes and downright bad behaviour. In May 1914, matters came to a head when Bishop was caught cheating in the examinations and was sent home for the summer holidays with the threat of certain dismissal hanging over him. But Bishop never returned to the College. In August 1914, war was declared and Bishop, along with many other of his fellow countrymen, left to fight for the Mother Country.

In June 1915, Bishop, as part of the contingent of the Seventh Canadian Mounted rifles, sailed for England on the troopship Caledonia, which docked at Plymouth on 23 June. While in England, stationed at Shorncliffe on the south coast, Bishop decided that he would prefer to fight his war in the air, and applied for pilot training in the Royal Flying Corps.

Bishop was interviewed for acceptance into the RFC by Lord Hugh Cecil, who told him that it would be six months or more before he could be accepted for training as a pilot, but that there was an immediate need for observers. This was better than nothing. Bishop applied, was accepted for training as an observer, and was posted to 21 Squadron, then working up to strength at Netheravon.

In January 1916, 21 Squadron left for France and was stationed at Boisdinghem. Bishop flew as an observer in the Squadron's RE7s and BE2cs for the next five months, until he was given leave on 21 May. A period of heavy drinking on the leave boat resulted in Bishop falling

William Avery Bishop with Nieuport 17 B1566.

down the gangplank when disembarking and severely twisting his knee, the same knee which had been injured in an earlier crash landing. Bishop ignored the pain, determined to let nothing spoil his leave, but at the end of his leave period the knee was worse and he was forced to report to the RFC hospital in Bryanston Square. An examination found that, in addition to his cracked knee, he had a seriously strained heart. He was graded unfit to return to France and active duty and taken into hospital for a period of rest.

After his recovery, Bishop was given a period of indefinite leave in Canada and on his return to England, in September 1916, he once more applied for pilot training. Again, he was unsuccessful; furthermore, it seemed that when he did return to France it would once more be as an observer. However, during his convalescence, Bishop had made a powerful friend in a Lady St.Helier, who had many connections in high places; strings were pulled, and in October 1916 Bishop was posted to the School of Military Aviation at Oxford to begin ground school training. With his previous experience at the front with 21 Squadron, Bishop easily passed the examinations in map reading and wireless, coming top of his class, and after a month he was posted to the Central Flying School at Upavon to begin his flying training.

Bishop had by now flown many hours as an observer and he anticipated no problems in learning to fly himself, but he soon found that the art eluded him. 'I tried very hard,' he wrote, 'but it seemed to me that I just could not get the proper "feel" of the machine.' In his anxiety to do well, Bishop sometimes gripped the controls too tightly, at other times too lightly, but finally, after three hours of dual instruction, his instructor sent him solo.

Bishop was very nervous, not at all sure that he was ready, but once in the air he felt better. He flew a couple of skidding turns. then prepared to land. He throttled back too far on his first approach and the engine almost stopped; he pushed the nose down, but too steeply. Bishop pulled the nose up a little, then down again, and in this fashion descended in a series of steps towards the ground. Finally pulling out, Bishop prepared to land. 'I leveled off and executed to perfection everything I had been told to do in order to make a perfect landing.' The only thing wrong was that the aeroplane was still some 40 feet off the ground. Bishop hastily put the nose of the machine down again and this time managed to land – from a height of eight feet.

Passing out from the CFS with three hours of solo flying, Bishop was posted to a home defense squadron at Northolt for an advanced flying course, including night flying. Here he ran foul of the commanding officer, who was all for dismissing Bishop from the course, but one of Bishop's instructors spoke up for him and earned Bishop a reprieve. As always when his plans were threatened, Bishop applied himself, and finally succeeded in winning his 'wings.'

After a further spell in England, stationed at Suttons Farm in Essex, Bishop was posted to 60 Squadron, based at Filescamp Farm, close by the village of Izel-le-Hameau. Bishop arrived at the Squadron on the evening of 9 March and reported to Captain Keith 'Grid' Caldwell, the 'C' Flight Commander, one of the most redoubtable fighter pilots of the war.

Bishop made an inauspicious start in 60 Squadron. He was still a very indifferent pilot and the sensitive little Nieuport 17, flown by 60 Squadron, was vastly different to the aeroplanes he had previously flown. Bishop began practice flights on 11 March and over the following two weeks he strained airframes, burst tires, and had at least one crash landing.

On his return from his first patrol over the lines, Bishop crashed on landing. Unfortunately, General Higgins, the 3rd Brigade Commander, was visiting 60 Squadron and witnessed the crash. Bishop was interviewed by the General, who was not impressed with his record and ordered that he should be returned to England for further training. This was a serious blow to Bishop's ambition to be a fighter pilot. He had nearly 20 hours of solo flying time and he knew full well that further training would not improve his flying; he would

Nieuports of 'C' Flight, 60 Squadron, at Izel-le-Hameau in early 1917.

always be a ham-fisted pilot. Things looked black, but 60 Squadron was short-handed and Major Scott, commanding the Squadron, told Bishop to wait until his replacement had arrived before leaving for England.

Bishop flew his second offensive patrol on 25 March. The four Nieuports of 'C' Flight crossed the lines between Arras and St.Léger and were attacked by three Albatros D.IIs. One Albatros overshot in its attack and pulled up directly in front of Bishop, filling his Aldis sight. Bishop fired a drum of ammunition into the Albatros and it went down vertically. Bishop followed, and as the enemy pilot pulled the Albatros out of its dive, fired again. Bishop reported that the enemy machine then crashed in a field to the north of St.Léger.

Bishop attempted to pull out of his dive but the engine of the Nieuport first refused to pick up, then stopped. Bishop now came under attack from gunfire from the ground. He had to get back over the front line and land as quickly as possible. He had a thousand feet in height and he began the long glide towards the British trenches. Bishop finally crossed the front line, but the ground ahead was full of shell holes and there was little clear ground in which to put the Nieuport down.

Ironically, Bishop made one of the best landings of his life; the Nieuport came to a stop among the shellholes, undamaged. Bishop spent the night with the infantry and the next morning decided to try to fly the Nieuport out. Taking off, the Nieuport bounced once, then twice, and staggered into the air. But just as Bishop thought he had succeeded, mud from the second bounce hit the propeller, smashing it. Bishop again spent the night with the infantry. The next afternoon a salvage party arrived from Filescamp Farm, loaded the Nieuport onto a trailer and took it and Bishop back to the Squadron.

When he arrived back, Bishop was fully expecting to be told that his replacement had now arrived and that he would be leaving for England, but to his delight he found that the Albatros had been awarded as a victory and that he had been given another chance. Bishop's biography relates that on the morning of Bishop's return, Major Scott asked him to lead a patrol that day, a Flight comprised of 'Grid' Caldwell – Bishop's Flight Commander – Lt. Horn, and Scott himself. The reason given for asking a pilot who had only flown over the front lines twice since joining the Squadron was that, being short-handed, the other pilots were tired. Contrary to what he told Scott, Bishop had had no sleep for 84 hours, but he took the Flight over the lines. No enemy aeroplanes were seen and on his return to Filescamp Farm Bishop slept for 16 hours.

Within a little over a month, Bishop had claimed another 11 victories, including a balloon. In April, he was promoted to Captain and given command of a Flight – an unprecedented appointment for a pilot who had been flying at the front for only just over a month. An award

William Bishop in the cockpit of a Bristol Scout.

Bishop during his command of 85 Squadron in 1918. The pilot on the left is the American Larry Callaghan.

of a Military Cross in May was followed by a DSO later the same month, and on 2 June 1917, Bishop took off on the raid that would win him a Victoria Cross.

In *The Courage of the Early Morning,* his biography of his father, Bishop's son writes that Captain Albert Ball, then serving with 56 Squadron, had flown to 60 Squadron on the evening of 6 May to ask Bishop to accompany him on an early morning attack on an enemy aerodrome. Why Ball should have done this is unclear. There were several experienced fighter pilots in 56 Squadron at the time, including 'Duke' Meintjes, who had flown with Ball when they were both in 60 Squadron in 1916. In addition, Ball was flying a patrol from 7:00p.m. to 8:40p.m. on the evening of 6 May and had earlier collected a replacement SE5 from the depot. Whether this meeting took place or not, Ball was dead the next evening, and Bishop went on leave the following day.

Bishop returned from leave on 21 May, and had claimed another three victories by the end of the month, bringing his total to 22, including 13 destroyed, with the remainder out of control.

Bishop had not forgotten the idea of an early morning attack on a German aerodrome. On 1 June he told Major Scott that he intended to carry out such a raid early the following morning. That evening Bishop asked to be called at 3:00a.m. the next day and went to bed early, missing a party in the Mess.

When Bishop left the ground in the early morning of 2 June, the weather was bad, low clouds and rain. By his own account, Bishop lost himself on his way to the lines and when he finally located an enemy aerodrome he found it deserted. Flying on, Bishop then saw another aerodrome, which he thought was either Esnes or Awoignt, although he had little idea of his actual position. There was a line of enemy machines in front of the hangars and Bishop made a pass across the field, firing at the parked machines. As he turned for his next pass, he saw an Albatros beginning to take off. Bishop attacked from 60 yards and shot it down. He then turned his attention to another and his attack so unnerved the enemy pilot that he crashed into a tree on the edge of the aerodrome. Two more Albatroses had now succeeded in getting into the air. One made for Bishop while the other stayed back. Bishop out-manoeuvred the Albatros pilot and shot him down to crash a few hundred feet from the aerodrome. The remaining Albatros now attacked Bishop, who had exhausted his Lewis gun drum and was attempting to replace it with a full drum. Finally getting the fresh drum in position, Bishop turned to face his attacker and fired a complete drum of ammunition at the Albatros. The enemy pilot turned away and dived for the safety of his own aerodrome.

Bishop then made for his base. He felt that the Lewis gun was now so much dead weight and, disconnecting it from the Foster mounting, threw it over the side of the Nieuport. When he finally re-crossed the front lines, Bishop was still unsure of his position and he landed to ascertain his exact whereabouts. Having done this, he flew on to Filescamp Farm, landing at half past five. He had been in the air for an hour and 43 minutes.

The next morning, Major Scott, commanding 60 Squadron, took both Bishop and Lieutenant William Fry MC by car to 3rd Army Headquarters. Scott and Bishop first saw Lord Dalmeny, Military Secretary to General H. Allenby, commanding the 3rd Army, before Bishop was taken in to see the General. This interview and Scott's report of Bishop's action on the morning of 2 June, resulted in the award of a Victoria Cross, gazetted on 11 August 1917.

It must be stated at this point that over the years considerable doubt has been expressed concerning the events of the morning of 2 June 1917 as related by Bishop. The main cause of this doubt is the lack of evidence in the German records of any attack on an aerodrome in the area of Bishop's operations, or any losses of men or machines that day which fit the circumstances. This book is not the place to review the argument surrounding the whole of Bishop's war service – in recent years it resulted in an inquiry in the Canadian Senate – but in the interests of historical fact it must be stated that a much admired aviation researcher, respected for both his thoroughness and fairness, the late Philip

85 Squadron's SE5as. The serial numbers have been censored.

Markham, carried out a detailed investigation into the events of 2 June. His conclusion was: 'There is not a shred of evidence to support Bishop's claims.'

By the end of July 1917, when 60 Squadron's Nieuports were replaced by the SE5a, Bishop had claimed a total of 36 victories. Before he left for England, on 16 August, he had claimed another 11 while flying the SE5a, bringing his total to 47.

Bishop attended an investiture at Buckingham Palace on his return to England and was awarded the VC and a Bar to his DSO. Before he left England in September, for leave in Canada, he had been promoted to Major.

Bishop returned to England in January 1918, bringing with him his bride of three months, Margaret. It had been intended that Bishop should be given command of the gunnery school at Loch Doon, but he was instead given command of 85 Squadron, a new fighter squadron, equipped with the SE5a and working up to readiness at Hounslow.

Under Bishop's command, 85 Squadron went to France on 22 May 1918, based firstly at Marquise, north of Boulogne, for three days, before moving to Petite Synthe, an aerodrome to the south of Dunkirk. Six days later, on 11 June, the Squadron moved again, to St.Omer. Although Bishop had by this time claimed another 12 victories and other pilots of the squadron had also claimed, it was not until its move to St.Omer that 85 squadron was officially flying on active duty.

Bishop recommenced his scoring on 15 June, with a Pfalz D.III, claimed as destroyed east of Estaires. Landing from a patrol on 16 June, Bishop was told that he was to relinquish command of 85 Squadron and was to be posted to Canadian Headquarters to help form the Canadian Air Force. Over the next three days, before he left the squadron, Bishop claimed another ten victories – on 19 June in appalling weather conditions he claimed five within ten minutes – bringing his total to 72, making him the highest scoring pilot in the RAF.

Again, it must be stated that **not one** of Bishop's victory claims can be substantiated by German losses. In 1987 the late Ed Ferko, the noted aviation historian on all matters pertaining to the *Luftstreitkräfte,* carried out a painstaking investigation into Bishop's victory claims in an attempt to match them to German losses. Ferko used the *Verlustliste der Deutschen Luftstreitkräfte im Weltkriege* (the official German list of aircrew killed in WW1); the records of *Kofl* 2, *Kofl* 4 and *Kofl* 6; and the unit histories of German squadrons operating on the same area of the Front as 60 and 85 Squadrons at the time in question. At the conclusion of his researches Ferko commented: 'It is not a pretty picture. This researcher has checked every possible German book, letter or record in his hands, looking for information relating to Bishop's claims – information either pro or con – nothing has been withheld which might confirm or deny any of Bishop's victories.' Ferko had failed to match a single victory claim made by Bishop against a known German loss for the day, time, and place in question.

After the war, Bishop returned to Canada. During the Second World War he served as an Air Marshal, with responsibilities for training. He died on 11 September 1956.

Major Lanoe George Hawker

Lanoe Hawker was born on 31 December 1890 at 'Homecroft' Longparish, Hampshire. Surviving pneumonia when only a few months old, he grew to be a serious and inquiring boy, small, but with 'a large head, high forehead, large, round eyes, and large ears.' At the age of 12, Hawker entered Stubbington House, a school on the Solent which prepared boys for entry into the Royal Navy. In 1905, nominated by Prince Louis of Battenberg, he sat and passed the exams for Britannia Naval College at Dartmouth, but he was not a robust child and the life at Dartmouth, particularly the sporting activities, into which Hawker threw himself with all his customary enthusiasm, were to prove too much for him. He caught a severe chill, which resulted in jaundice. He refused to report sick until it was too late and the result was a strained heart. Hawker gave up the idea of a career in the Royal Navy and it was while still convalescing at the new family home in Herefordshire that he began to experiment with kites and model aeroplanes.

Lanoe George Hawker.

Having recovered his health, Hawker decided to make the Royal Engineers his career and he began to study for the Royal Military Academy at Woolwich, known as the 'Shop.' The family rented a house in Cricklewood, north London, and engaged a live-in tutor for both Lanoe and his younger brother, Tyrrel. While at Cricklewood the boys saw a short cinema film of the Wright aeroplane in flight and during the summer of 1910 both boys were taken to their first aviation meeting, at Bournemouth. From this moment on Lanoe Hawker was fired with enthusiasm for all things concerned with aviation and the brothers spent all their spare time in designing and building model aeroplanes, Lanoe consulting every book he could find on their theory and construction.

Hawker entered the 'Shop' in 1910. His great friends there were Gordon Bayly, Ian Gourlay, and David McMullen, all four sharing a common passion for aviation.

In 1910 Lanoe and Tyrrel joined the Royal Aero Club, which entitled them to free entry to the flying meetings then being held at Hendon, only a short distance from Cricklewood. Nearly every weekend was spent at Hendon and it was there that the future VC had his first flight, in a Valkyrie, piloted by a Mr. Barber. The flight inspired Hawker to learn to fly and after an abortive start – Mr. Barber's flying school closed because of the inherent instability of the Valkyrie, which had killed an Army pilot under instruction – he enrolled in the Deperdussin school at Hendon. On 4 March 1913, both Lanoe and his friend McMullen passed their tests at Hendon and were awarded their Royal Aero Club certificates. A fortnight later, Hawker's other great friend, Gordon Bayly, also passed.

The three friends had by now passed out at the 'Shop,' were commissioned in the Royal Engineers, and were at the School of Engineering at Chatham. While at the school, Hawker owned a G N Sports car, which had a belt drive and a 'brutal gear change.' Hawker, with his innovative engineering mind, devised a differential belt drive which made a gradual gear change possible through a system of pulleys. Hawker took out a patent on this device, but when it proved impracticable because of the strain on the belts, he turned his mind to an even more ingenious solution. A hydraulic motor on a sliding bed, driven by the car's engine, piped liquid to hydraulic rotary motors fixed to each wheel of the car. The car could be driven without the need for a clutch, gearbox or differential gears. In addition, restricting the flow of the liquid to each wheel gave highly efficient synchronized braking. This highly promising invention was shelved at the outbreak of war in 1914.

In June 1914 Lieutenant Lanoe Hawker received orders that he was to report to the Central Flying School in September, but on 30 July a telegram came: 'Report at once at Central Flying School.' Five days later Britain was at war with Germany.

Lanoe Hawker arrived at the Central Flying School as

Bristol Scout 1611. Hawker won the VC for his action on 25 July 1915 while flying this aeroplane.

one of a batch of pupils replacing those who had been posted to squadrons, one of whom was Gordon Bayly, who had gone to France with 5 Squadron. Hawker soloed on 11 August, but he was already impatient with the slow pace of his training. Aeroplanes were few and were constantly being crashed by pupils less adept. Hawker was eager to follow Bayly to France, but on the 22 August Bayly failed to return from a reconnaissance, and he and his observer were reported missing – the first Royal Flying Corps crew to be lost in the war. They came down behind the German lines and were the first firm indication the German High Command had that the British Expeditionary Force was now in France. Hawker was devastated by the loss of his friend, but even more eager to get to France, and on 3 October 1914 he finally passed out of the CFS and was posted to 6 Squadron.

On 7 October 1914, 6 Squadron flew their eight BE and four Henry Farman aeroplanes to Belgium, landing at Bruges. The squadron flew its first reconnaissance the very next morning, at 7:40a.m., with Hawker in the observer's cockpit of BE2a 492, piloted by Capt. A.C.E. Marsh. That afternoon the Squadron moved to Ostend, but the German army was nearing the town and the Squadron moved again on 12 October, to St.Pol. The day of the move to St.Pol was foggy, and at one time it looked as if the Squadron's aeroplanes would have to be burnt to deny them to the enemy forces, but Hawker led the Squadron along the coast in his Farman, finding his way by the waves breaking on the beach, and the squadron landed safely at St.Pol.

Hawker, annoyed at having no means of retaliation against the groundfire he was experiencing on his reconnaissances, took to carrying bombs and a revolver aloft with him, and on the last day of October 1914 had his first encounter with an enemy aeroplane.

'1.11.14

I met a German biplane yesterday afternoon and fired six shots at him with my revolver – no good, of course, but he took fright and made for home.'

Despite such diversions, Hawker was bored. Bad weather, with no flying possible, plus his machine being unserviceable, all contributed: on 8 November he wrote: 'I am having a very dull time just now, while my machine is being tuned up.' And two days later, in another letter home. 'My luck is out and I'm having a horrid time. My machine was finished yesterday and I hoped my days of boresome inaction were over – there is nothing so tiring as being bored – the weather was horrid, wind and rain, but I took it out for a short trip to see if all was OK. It showed promise of being quite a success but required some minor adjustments. This morning, however, it was blowing a gale so I couldn't try it again and while I was down at lunch the brute broke loose and was completely wrecked. So I'm worse off than before, which is most annoying as I've been bored to tears this last week and was looking forward to getting into harness again.'

There had been few opportunities for air combat during 1914, but as the spring of 1915 passed into summer, and the aeroplanes of both sides became armed – although many in makeshift ways – it was realized that it was imperative to stop enemy aeroplanes from carrying out their now vital reconnaissance work. Hawker's main duties with 6 Squadron were still those of reconnaissance, artillery ranging, and photography, but he soon began to attack any enemy aeroplane seen during these routine duties, and on 2 March 1915, while flying a routine reconnaissance, Hawker saw British anti-aircraft shells bursting around an Aviatik to the south of Hazebrouck. When Hawker finally closed with the enemy, his BE2c 1780 was at 7,000 feet, two thousand feet lower than the Aviatik, but Hawker fired four shots from his revolver. These had no effect. Hawker's observer, Cherry, was flying in the front cockpit of the BE, the usual arrangement in those days, and could not bring his rifle to bear because the proximity of the top wing

The gun mounting on Bristol Scout 1611.

obscured his view of the Aviatik. Hawker took the rifle from Cherry: '...leave the machine to look after itself, good old "Stability Jane",' he later wrote, 'and pot 3 at it myself.' The Aviatik pilot dived for his aerodrome at Douai, joined by another Aviatik. 'Both making for home, they're no sportsmen.'

Hawker had now given the rifle back to his observer and Cherry kept up a steady fire on the first Aviatik, which shut off its engine and began to glide down. Cherry switched his attentions to the second enemy machine, but Hawker was conscious of the strong east wind blowing, which would make return difficult and finally turned away having driven down both Aviatiks: 'put the wind up them, spoilt their reconnaissance, saved HQ at Hazebrouck from bombs, and drove them ingloriously to earth. Had a long fight back against the wind and continued our reconnaissance.' This fight was mentioned by Sir John French in his bi-weekly report.

On 8 March, 6 Squadron moved to Poperinghe aerodrome, sharing the field with the French *Escadrille* MF 33, and air activity began to increase as the weather improved. Hawker flew almost every day. In addition to his usual duties he dropped propaganda leaflets over the German lines and packets for agents working in enemy territory. On 29 March, while ranging for the artillery, Hawker switched off the engine of his BE to discuss the ranging with his observer. The BE was still over the lines and a shell burst just above the aeroplane, 'just above and behind my right ear! Nasty jar, bits whizzing by and shadow over machine, put wind up me properly, switched on and fled at 100mph.' Hawker suffered from slight concussion from this incident, he was almost deaf in his right ear; both ears were very sore and he suffered constant headaches for some weeks.

Hawker had seven days leave at the beginning of April 1915. His return to the Squadron coincided with a spell of stormy weather, but this past he threw himself once more into the daily round of squadron work. By this time Hawker was recognized as one of the most experienced and reliable pilots in the RFC and was given missions of special importance. On 18 April, 6 Squadron was given the task of attacking the Zeppelin shed at Gontrode and Major Gordon Shephard, commanding the squadron, selected Hawker as the best pilot to carry out the raid.

Hawker took off carrying three bombs. On arrival over the enemy base he found that the shed was guarded by a tethered observation balloon carrying a machine gun. Hawker dropped his first two bombs from 4,000 feet but as these had no effect – one burst 200 yards short of the shed and the other either failed to explode or Hawker failed to see it – he decided to deal with the 'silly captive sausage balloon' so that he could safely bomb from a lower altitude. He spiraled down on the balloon, throwing hand grenades at it until he was below it. He then switched off his engine, dived across the Zeppelin shed and dropped his third bomb from about 200 feet. To Hawker's disappointment, this bomb hit about six to ten feet short of the shed. Switching on his engine, Hawker flew away, keeping low to confuse the enemy anti-aircraft fire. The gunner in the observation balloon had made good shooting and had hit Hawker's aeroplane several times, one bullet hitting the rubber joints of the induction pipe, causing the engine to run badly. For this raid, and his other work, Hawker was awarded a DSO and was promoted to the command of 'A' Flight.

On 24 April, during the Second Battle of Ypres, Hawker was wounded in the foot by ground fire, the bullet entering his heel at the Achilles tendon, striking the bone but not breaking it. Although in considerable pain, Hawker finished his reconnaissance before returning to base. After making his report, he had the wound dressed and 'I hobbled across to the mess and had supper.' Despite his wound, which made walking difficult, Hawker flew throughout the battle, at times being carried both to and from his aeroplane, and was congratulated by the Wing Commander, Colonel Burke, for his 'splendid work.'

At the end of April the crisis in the ground fighting was past and Hawker was given sick leave, which was extended to three weeks when it was discovered that his wound was more serious than it appeared. Tyrrel was also on leave and was struck by the change in his brother. 'Gone were the soft boyish curves of his mouth...His lips were set in a firm line from the constant habit of drilling his spirit to face danger unflinchingly. His large bright eyes had lost their gay laughter...had sunk back into their dark surrounds.'

On 20 May Hawker returned to France. Burke wrote: 'He is now back, flying as daringly as ever...' and on June 3 he was sent to the Aeroplane Depot at St.Omer to collect a single-seater Bristol Scout C, No.1611. At this time the small, more manoeuvrable single-seat aeroplanes which the RFC had on charge, machines such as the Martinsyde S.1 and the Bristol Scout Type C, were being armed with a variety of weapons and flown by those pilots in the squadrons who had shown fighting spirit and aggressiveness. With the help of his mechanic, E.J. Elton, Hawker devised a mounting for a Lewis gun on the Bristol. The synchronizing gear, which would enable the gun to be fired through the propeller without damaging it and which would allow the aeroplane to be aimed as a gun, had not yet been perfected, and Hawker's solution to avoid shooting off the propeller of the Bristol was to position the Lewis gun to fire forwards at an angle, missing the arc of the propeller. The difficulties in aiming the Lewis were considerable but Hawker was delighted with his new aeroplane. 'I have a beautiful little toy, a new Bristol Scout that goes at 80 and climbs 5 or 600 feet a minute!'

At first Hawker gained no victories in the Bristol, although he had a number of combats and sent down several enemy aeroplanes with smoke issuing from them, but on patrol on the evening of 25 July he saw two enemy aeroplanes over the Salient. Hawker lost no time in attacking the nearest. The German aeroplane was also armed with a machine gun and scored several hits on the Bristol, but Hawker emptied a whole drum into it and another into its companion. The enemy aeroplanes both dived away. Half an hour later Hawker saw another hostile two-seater 10,000 feet above Hooge. Hawker manoeuvred into the eye of the evening sun and attacked, opening fire from 100 yards. The enemy biplane went down out of control, then burst into flames. The wreckage fell in the British lines and a map was found on the body of the dead observer, not only showing that he had successfully pinpointed the position of a British battery, but also giving the map reference of a particularly troublesome German battery, known to the British artillery as 'Percy,' which had eluded the efforts of the RFC to locate for sometime.

A British anti-aircraft section later reported that one of the enemy biplanes first attacked had also been forced down, in its own lines, and the Army Commander, General Plumer, sent a telegram to 2nd Wing HQ. 'The Army Commander is very glad to hear of your success this evening and wishes his congratulations to be conveyed to the officer concerned.' For this action, and for his untiring and magnificent work during the whole of his time in France, Hawker was awarded the Victoria Cross, gazetted on 24 August 1915. It was the first VC to be awarded for air combat between aeroplanes.

Hawker continued his constant round of reconnaissance in one of the FE2a pusher biplanes, a new type which the squadron was now receiving, and solo patrols in the Bristol Scout. On 11 August, returning from a reconnaissance in FE2a 4227, Hawker was attacked by one of new Fokker E-type monoplanes These Fokkers were armed with a machine gun synchronized to fire through the propeller and the fast little monoplane attacked the FE at 9,000 feet between Lille and Roubaix. Hawker and his observer, Lt. Noel Clinton, had already had three combats during the patrol – one lasting 20 minutes – and had only a drum and a quarter of ammunition left when Clinton saw the Fokker swooping down from behind the FE. Hawker immediately turned the heavy FE to face the Fokker and Clinton fired a burst into it from 50 yards range. The German pilot dived his

Precautions against the cold – Lanoe Hawker well wrapped up for flying in the winter of 1915.

The DH2.

machine 'absolutely vertically' for almost 4,000 feet. He then flattened out, but crashed on the outskirts of Lille.

Despite these successes, Hawker was now almost at the end of his endurance. He was overworked, both in the air and on the ground. Gordon Shephard could ill afford to spare him, but the signs of strain were obvious and he sent Hawker home on leave. Hawker returned to France on 28 August, but it was still plain that the strain of nearly a year's war flying had taken its toll and he was posted to Home Establishment on 20 September 1915. Hawker was the last of the original pilots who had flown to France with 6 Squadron in October 1914; although he deeply regretted having to leave the squadron he knew that he was at the end of his strength. 'I'm not quite as well as I should like,' he had admitted on his return to France, 'and if I don't get better soon I shall apply for a home job.'

On 28 September 1915, Hawker was ordered to report to Hounslow to take command of the newly-formed 24 Squadron, destined to be the RFC's first fighter squadron. On 10 January 1916 the squadron received its first DH2 and by the end of the month was up to its full complement of 12. The DH2 was a single-seat biplane, armed with a single Lewis gun, its pusher configuration negating the necessity of a gun synchronization gear, and it was hoped that its arrival in France would counter the German Fokker E.III monoplane which was taking such a heavy toll of the RFC machines. On 8 February 1916, 24 Squadron flew to St.Omer in France and came under the orders of RFC Headquarters.

As Squadron Commander, Hawker now had no duty to fly patrols but he tested many of the replacement DH2s over the following months and his active and inventive brain was constantly at work to improve the performance of his squadron. Hawker's combat tactics were simple. A board was erected on which a single sheet of paper was headed.

'Tactical Orders by O.C. No.24 Squadron, Royal Flying Corps.'

Under this were only two words 'Attack everything.' It was signed by Hawker.

On 23 November 1916, three DH2s of 24 Squadron set off from their base at Bertangles. Their orders were to fly a defensive patrol over the 4th Army Front. At 1:50p.m. the DH2 pilots spotted a pair of Roland two-seaters flying at 6,000 feet north east of Bapaume. The leader of the DH2s, Capt. J.O. Andrews, attacked the Rolands, sending them diving east, but he then saw two large patrols of enemy fighters approaching with the advantage of height and pulled away. As he did so, Hawker, flying DH2 No.5964, dived past him in pursuit of the Rolands. Andrews and his companion, Lt. R.H.M.S. Saundby, were immediately attacked by the hostile formations and lost sight of Hawker in the general fight. Andrew's DH2 was

24 Squadron at Bertangles in the summer of 1916. The red and white interplane strut markings used by the Squadron can be clearly seen. A solitary Morane Bullet is on the extrement right.

hit in the petrol tank and he turned for the lines. His last sight of Hawker was of his commanding officer in combat with a German fighter.

Hawker was fighting Manfred von Richthofen, flying an Albatros D.II, a vastly superior aeroplane to the DH2. After a long fight, lasting some minutes, Hawker must have realized that the prevailing wind was pushing him further into enemy territory and he broke off and dived for the safety of the British Lines. Richthofen followed and his fire hit Hawker in the head. The DH2 crashed near the ruins of Luisenhof Farm, just south of Bapaume on the road to Flers. Lanoe Hawker was dead. Manfred von Richthofen was to score 80 victories before his death in April 1918. Hawker was the 11th victory of the emerging German ace.

Lieutenant Gilbert Stuart Martin Insall

The Vickers FB5 – popularly known as the Vickers Gunbus – was a two-seater pusher aeroplane, a steady improvement of the type which had been under development since 1912, when the Vickers Company had been awarded an Admiralty contract for a fighting aeroplane armed with a machine gun. The first FB5 to reach France was with 2 Squadron in February 1915, but it was not until late July 1915 that 11 Squadron, fully equipped with the Gunbus, commenced operations over the 3rd Army front.

Although 11 Squadron can lay claim to having been the first two-seater fighter squadron to be completely equipped with one type of aeroplane on the strength of the Royal Flying Corps in France, its duties were still varied and the squadron was called upon to carry out many tasks: offensive and defensive patrols; escorts to bomber and reconnaissance aeroplanes; the taking of both vertical and oblique photographs; distant and close reconnaissance; and 'when we hadn't anything else on hand,' a little artillery observation. In the event the effectiveness of the Vickers as a fighting aeroplane was to be short lived. With the arrival of the Fokker E type monoplanes over the Western Front in the summer of 1915, the supremacy of the air once more swung in favour of the German air service.

When No.11 Squadron arrived in France, stationed at Vert Galant, it was commanded by Major G.W.P. Dawes. Dawes had three excellent Flight Commanders, all veterans: Captains L.W.B. Rees, R.H.L. 'Pip' Playfair, and C.C. Darley. Also serving in the squadron were two brothers, Gilbert Stuart Martin Insall, a pilot, and his younger brother Algernon J. Insall, who was an observer.

Gilbert Insall was born in Paris on 14 May 1894, the son of a doctor practicing in the French capital, and the two Insall brothers were studying at the University of Paris when it became obvious that war with Germany was only a matter of time. On 2 August 1914, when news broke that German troops had invaded the Duchy of Luxembourg and had crossed the border into France, their father lost no time in closing his practice and house, and on the morning of 4 August 1914 the Insall family boarded the cross channel steamer Maid of Kent for England.

In September 1914 the Insall brothers joined the University and Public Schools Brigade of the Royal Fusiliers and were in training until the early spring of 1915. On parade one morning an urgent appeal from the War Office was read out asking for volunteers to train as pilots in the Royal Flying Corps. Neither paid any attention at first, despite having flown with Maurice Farman while they were living in France, but their platoon commander 'immediately pounced upon us.' He had remembered the flying 'experience' of the brothers. They could think of nothing better to say than 'Why not,' and that evening filled in the forms to be transferred to the RFC. Against the question 'Previous experience' they wrote 'Close interest in progress of aviation while resident on the continent (Paris) Passenger ascents at Buc (Versailles) with M. Maurice Farman.' A week or two later they received orders to report to

Gilbert Stuart Martin Insall.

The Vickers FB5 – 'The Gunbus.'

Brooklands as probationary 2nd Lieutenants in the Royal Flying Corps.

The Insall brothers took their Royal Aero Club certificates only a day apart – Algernon's being 1109 and Gilbert's 1110 – and were then posted to Netheravon. It was while they were at Netheravon that Algernon had a bad crash in a Farman. This affected him so badly that each time he came in to land he was unable to control the aeroplane and the instructor had to take over. This temporary loss of nerve may have been overcome in time, but 11 Squadron, working up at Netheravon, were to leave for France as soon as it had received its full complement of Vickers FB5s and the brothers had no wish to be separated. Algernon re-mustered as an observer and went to France with the squadron.

On 7 November 1915, Gilbert took off from Bertangles in Vickers Gunbus No.5074 on a special reconnaissance. His observer was a young Scot, First Class Air Mechanic Corporal T.H. Donald. As they neared the front lines, Gilbert saw an observation balloon and dived to attack it. The attack was unsuccessful and as he pulled away he saw an Aviatik two-seater approaching from the north. The hostile machine was 2,000 feet higher than the Vickers and was flying down the line of trenches. Insall turned to give Donald a shot at the enemy with his Lewis but the Vickers was only just within range and the Aviatik immediately turned east. Insall continued to climb, to give the impression that he was not going to press home the attack. The ruse succeeded. The enemy pilot, seeing that the Vickers was not following him, turned back and followed the British machine. As the enemy aeroplane came within range, Insall turned towards it and Donald opened fire. The enemy pilot quickly turned and dived for the protection of the anti-aircraft battery at the base of the balloon. Undeterred, Insall dived after it and Donald fired a complete drum of Lewis into the fleeing two-seater, stopping the engine. The German observer returned fire until both contestants flew into a cloud and lost sight of their opponent. On emerging from the cloud Insall and Donald saw the German aeroplane landing in a plowed field, the pilot and observer scrambling from their machine and running for cover, the observer carrying his machine gun.

Insall turned the Vickers, flew over the two-seater, and released the incendiary bomb he had earlier intended for the balloon. The bomb burst close to the Aviatik and as the Vickers climbed away the enemy machine was enveloped in a cloud of thick smoke. The Vickers was by now very low and over the German side of the lines. Troops in the neighbouring village began to open fire, joined by the balloon's anti-aircraft defenses, and as the little pusher fled for the safety of the British Lines it ran the gauntlet of hundreds of guns. Reaching the German front-line trenches, the fire increased, round after round hitting the wings and tailplane. As the Vickers reached No-Man's Land a bullet hit the petrol tank and the engine stopped. Insall put the nose of the Vickers down, managed to clear a small wood, and landed on the far side of the trees, only 500 yards from the front-line trenches, near the village of Agny, south of Arras.

The German artillery began to shell the Gunbus, aided by a spotter aeroplane, but as darkness fell it was still undamaged and on the arrival of a party of groundcrew from Bertangles, under the command of Captain L.W.B. Rees, it was moved further back, a tarpaulin screen erected around it and work was begun on the fitting of a new petrol tank. As dawn broke attempts to start the engine were unsuccessful and it was found that the replacement petrol tank was not holding its pressure. Working feverishly behind the screen, the groundcrew located the source of the leak. It was sealed and the engine started. Expecting a rain of shells at any moment, Insall had no wish to endanger Donald's life in a hazardous takeoff and he had the front cockpit filled with sandbags. There was no time to take the precaution of warming up the Monosoupape engine of the Gunbus and Insall risked an immediate takeoff. Watched by hundreds of French soldiers, who were holding the allied front line at this point, the little Vickers lurched over the uneven ground. To the cheers of the French Poilu it staggered into the air. Insall could not resist the temptation to show the enemy that he had escaped their intentions and was careful to show himself to the

A Gunbus in enemy hands. This FB5, serial 1621 from 16 Squadron, was forced down intact on 28 February 1915. The crew, 2Lt. M.R. Chidson and 2Lt. D.C.W. Sanders, were uninjured and taken prisoner.

German front-line trenches. This piece of effrontery drew rifle fire of such intensity that Insall's well wishers had to throw themselves on the ground as a hail of bullets came through the trees. Once in the air and a course set for Bertangles, Insall saw that his engine had been hit by two bullets, one of which had bent an exhaust valve tappet rod. However, the engine continued to run and Insall landed safely at his base.

For this action, Insall was awarded the Victoria Cross, gazetted on 23 December 1915. Donald was awarded the Distinguished Conduct Medal.

Five weeks later, while on patrol, again with Donald, Insall attacked an enemy two-seater whose return fire hit and stopped the engine of the Vickers and wounded Donald in the leg. The fight had taken place well behind the German lines and although there was little hope of being able to reach friendly territory, Insall put his aeroplane in as shallow a glide as possible. Almost at once the Vickers came under intense anti-aircraft fire and a shell exploded just below it, the base of the shell wounding Insall in his left buttock. Insall lost consciousness for a moment but recovered and managed to land near Achiet.

Insall and Donald immediately attempted to set fire to the Vickers. In spite of his leg wound, Donald kicked the incendiary bomb out of his housing below his seat and passed it back to Insall. But the bomb was jammed in its tube container and before Insall could free it a German NCO reached the scene and covered him with a pistol. Both Insall and Donald were prisoners of war.

After recovering from his wound, Insall was imprisoned in Heidelburg POW camp and escaped in 1916, only to be recaptured five days later. He was then sent to another camp, at Crefeld, but again he managed to escape, this time through a tunnel. He was yet again recaptured and this time given a term of solitary confinement in Strohen camp as a punishment. Insall escaped from Strohen on 28 August 1917, with two companions. This time he evaded recapture, and after a journey of nine days crossed the Dutch border to freedom. Insall received a Military Cross for his escape attempts and served throughout the rest of the war, becoming a highly skilled Sopwith Camel pilot.

Gilbert Insall remained in the RAF after peace was declared in 1918, served throughout the inter-war years, and in WW2 was at one time in command of RAF Uxbridge. He finally retired from the service on 30 July 1945 with the rank of Group Captain. Living in Yorkshire, Insall nearly died of a heart attack in 1969 when his home was burgled and all his decorations, including his Victoria Cross, were stolen. Luckily the medals were recovered in 1970. Insall died on 17 February 1972.

Lieutenant Alan Jerrard

Alan Jerrard has two claims to fame: he was the only pilot to win a Victoria Cross on the Italian Front and he was the only Sopwith Camel pilot to win Britain's highest award for bravery. Jerrard was born in Lewisham, London, on 3 December 1897. He was educated at Oundle and Birmingham University. On 2 Jan. 1916, Jerrard was commissioned as a 2nd Lieutenant in the South Staffordshire Regiment, but on 23 September he was seconded to the Royal Flying Corps to begin flying training and he passed out from the Central Flying School at Upavon on 14 June 1917 with an above average assessment as a pilot.

Jerrard's first posting was to 20 Squadron, to gain experience in flying the RE8, but his exceptional qualities as a pilot convinced the authorities that he would make a good fighter pilot, and on 4 July 1917 he was posted to London Colney in Hertfordshire to train on the SPAD VII. On 24 July Jerrard was posted to France, joining 19 Squadron – a SPAD squadron – at Estrée Blanche, and flew his first patrol over the lines five days later. He became lost during this patrol and force landed at St.Omer.

Alan Jerrard.

19 Squadron did no war flying for the next few days and it was not until the evening of 5 August that Jerrard flew his second patrol, led by Captain John Leacroft, an experienced Flight Commander with two victories to his credit. Leacroft had served with the RFC in Egypt as an observer before training as a pilot and he would end the war with more than 20 victories.

On the evening of 5 August, Leacroft led four SPADs – one had dropped out with engine trouble – across the lines and attacked ten enemy aeroplanes, chasing them down without achieving a decisive result. Jerrard lost the formation in the dive, Leacroft diving away too quickly for him to follow, and he continued down to ground level, under the mist, looking for ground targets. He found a convoy four miles long and shot it up, leaving several vehicles in flames. The ground mist then thickened and Jerrard climbed back to 10,000 feet, but the engine of the SPAD then cut out and he glided down to crash into the side of a railway embankment. Troops had to dig Jerrard out of the wreckage: he had a broken nose, his jaw was broken in three places, and he was cut and bruised.

Jerrard was invalided back to England and spent several months in hospital. On his recovery, and after a refresher course flying rotary-engined aircraft, Jerrard was posted to 66 Squadron in Italy. Major J.T. Whittaker MC was then commanding 66 Squadron, one of three Sopwith Camel squadrons on the Italian Front, and Jerrard reported to his new CO on 22 February 1918.

Jerrard opened his score with 66 Squadron with a Berg, shot down out of control over Vittorio five days after reporting. He followed this victory with a balloon on 7 March, another Berg on 11 March, and an Albatros claimed on 21 March. Three more Albatros scouts were awarded to Jerrard on 30 March, all in the vicinity of Mansue aerodrome. These last three victories were awarded to Jerrard in his absence, after he had been taken prisoner of war, all three being from his last fight in Italy.

On 30 March 1918, Captain P. Carpenter led three Camels of 66 Squadron on a No.12 Offensive Patrol, a designation which left the type of any attack to the discretion of the Flight Commander. Peter Carpenter had scored five victories in France with 45 Squadron before going to Italy with the Squadron in the winter of 1917, and he had been posted to 66 Squadron as a Flight Commander in January 1918, adding another seven victories to his score by the morning of 30 March. The Flight, consisting of Carpenter in Camel B.7387, Lieutenant H.R. Eycott-Martin in B.7283, and Alan Jerrard in B.5648 marked 'E,' took off from their aerodrome at San Pietro-in-Gu and made for their general patrol area, climbing steadily to 13,000 feet.

At 11:35a.m. the three Camels sighted a formation of four Albatros D.IIIs escorting a Rumpler at 14,000 feet over Fontane. The enemy formation was crossing the Allied lines back into their own territory. The three Camels could not close the height sufficiently quickly to be able to attack and Carpenter led them into the eye of the morning sun, climbing hard. Carpenter then saw that the enemy formation was losing height and that the Rumpler was about to land on its aerodrome at Mansue, the base of *Flik* 32D, and he led the Camels down in an attack.

From this point, events become confused. According to the 66 Squadron Combat Report – which was not countersigned by either Carpenter or Eycott-Martin on their return, but by Major Whittaker – six Albatros D.IIIs were destroyed in the fight which followed: two to Eycott-Martin; one to Carpenter and three to Jerrard.

The report states that Carpenter shot down the first of the Albatroses, about three miles south of Mansue, Jerrard following his leader down and destroying another, forcing the Albatros down to within a hundred feet of the ground until it crashed in the same place as Carpenter's victory. Eycott-Martin confirmed Carpenter's victory and Carpenter and Eycott-Martin both confirmed Jerrard's. Carpenter and Eycott-Martin were at 6,000 feet and they now saw Jerrard, who had stayed low after forcing down the Albatros, attacking the enemy aerodrome at Mansue from 50 feet, strafing the field and attacking any enemy machines attempting to take off. The Combat Report states that at this time there were 19 enemy machines in the air and it appears that six of these – Albatros D.IIIs – attacked Jerrard. Eycott-Martin

Alan Jerrard sits dazed and disconsolate by his crashed Camel.

dived to Jerrard's aid and shot an Albatros off his tail, which crashed in a field next to the aerodrome. This was confirmed by Carpenter. Jerrard then shot down another Albatros, which crashed in the same field, and this was also confirmed by Carpenter. Jerrard then returned to his attacks on the aerodrome.

Carpenter was by now fighting three Albatroses and lost sight of his companions. Eycott-Martin was attacked from below by an Albatros D.III, but Jerrard shot it off his tail and it crashed two fields north of the aerodrome, confirmed by Carpenter and Eycott-Martin. Eycott-Martin then attacked another Albatros, forcing it down without firing a shot, and it was seen to crash by Carpenter and himself. Jerrard was now fighting – 'attacking vigourously (sic)' is the wording of the Combat Report – with six Albatros D.IIIs. Carpenter and Eycott-Martin attempted to go to his aid but were driven off by other enemy machines and forced to break off the action, being heavily outnumbered.

Jerrard, it then appears, was able to escape from the six Albatroses he had been fighting and he rejoined his companions. The Combat Report is very confusing here – as indeed it is in many places. After stating that on rejoining the formation Jerrard was 'flying very weakly as though wounded,' it goes on to say: 'Ten D.IIIs were following Lt. Jerrard. Capt. Carpenter and Lt. Martin were unable to go to his assistance owing to other EA and observed Lt. Jerrard, who although apparently wounded, repeatedly attacked these EA until he crashed four miles west of Mansue Aerodrome.'

If Jerrard had rejoined the formation, why was it stated that the ten D.IIIs were following him alone; and why were Carpenter and Eycott-Martin unable to go to Jerrard's assistance if they were flying in formation with him. How could Carpenter and Eycott-Martin tell that Jerrard was wounded? 'Flying very weakly' is a very imprecise term; what exactly does it mean? Yet the very next sentence of the Combat Report states that Jerrard, 'although apparently wounded' – which he was not – 'repeatedly attacked these EA.' Group Captain Norman Macmillan, in his book *Offensive Patrol The Story of the RNAS, RFC and RAF in Italy 1917–1918,*[1] gives the opinion that the anomalies in the Combat Report arise from faulty transcriptions of verbal statements made by Carpenter and Eycott-Martin while they were still stressed by their recent combat, compounded by having lost Jerrard. Neither countersigned the report and it is more than likely that they never read it before it was forwarded to Wing. It was on the basis of this report, despite the many anomalies, that Jerrard was awarded a Victoria Cross for his part in the action on the morning of 30 March. As Macmillan commented: 'Before this supreme award is made the initial recommendation passes through many upward stages, at each of which it is scrutinized. The anomalies in the report and its

1. Jarrolds. London, 1973.

Jerrard leaving the scene of his crash with Benno Fiala.

defective language construction must have been disregarded by concentration on the salient factor of Jerrard's lone swoop "into the jaws of death."'[2]

The Austrian account of the events on the morning of 30 March is very different. The four Albatros D.IIIs escorting the Rumpler, seen by the Camel pilots, were led by *Oberleutnant* Benno Fiala *Ritter* von Fernbrugg, commander of *Flik* 51J, flying Albatros D.III No.153.155. After seeing the Rumpler safely to its base at Mansue, the Albatroses had turned towards their own base at Ghirano when they were attacked by three Camels, coming down on them out of the sun from a great height. This agrees with the 66 Squadron Combat Report. The Camel attacking Fiala was then attacked by *Offizierstellvertreter* Stefan Fejes, who forced it to break away from his leader and dive away to the south, towards the River Piave. Seeing another Camel flying towards him, Fiala opened fire from long range. The Camel dived a little, then turned towards the Italian lines. Turning, Fiala got onto the tail of the Camel and fired another 100 rounds into it. He was joined in this attack by *Stabsfeldwebel* Bönsch. The Austrian pilots then saw that the Camel's engine had stopped and it glided down to crash in a tree-lined meadow at Gorgo al Montico, two miles south of the enemy aerodrome at Mansue. The Camel hit one of the trees in landing, tearing off its port wings and twisting the rear of the fuselage and tailplane round before coming to rest on its nose. Despite the severity of the crash, Jerrard was relatively unharmed and was extricated from the wrecked Camel by Austrian infantry.

Landing at Ghirano, about four kilometers away from the crashed Camel, Fiala went by car to the site of the crash, took charge of Jerrard, and drove the still dazed pilot to the nearest Army Headquarters at Oderzo.

On the evening of 30 March, the Headquarters of General Diaz, the Italian Commander in Chief, reported that six Austrian machines had been shot down in the vicinity of Mansue. The Austrians reacted quickly to this, denying the loss of six aircraft and calling the final paragraph of the Diaz report 'over exuberant.' The Austrians claimed that two Camels had been shot down in the action: one by Fiala and another by *Feldwebel* Stefan Fejes, whose victory was confirmed by witnesses on the ground. Yet only one Camel was lost – Jerrard.

It has been said that truth is the first casualty of war and each side put forward their own version of the events of 30 March. The Austrians admitted only one casualty: *Feldwebel* Fejes of *Flik* 51J, who was slightly wounded in the foot; and although the 66 Squadron Combat report gives the fight as being of some 25 minutes duration, in the official Austrian report of the action the impression is given that the fight with the Camels was very brief. Yet the Austrians admitted that three aircraft of *Flik* 51J were so badly damaged in the fight that they had to be returned to the *Fliegerstappenpark* (Aircraft Repair Park), which hardly argues a brief combat.

Macmillan interviewed Alan Jerrard in 1963 in an attempt to clarify some of the inconsistencies in the account of the fight which led to the award of his VC. Jerrard was then 66 years of age, but Macmillan wrote that he looked over 70. Questioned, Jerrard stated that he had attacked the aerodrome at Mansue – which he had not admitted when interrogated by the Austrians after his capture – but could throw little new light on the subject. Perhaps he was embarrassed by his award, given for an action which he knew differed substantially from the account given in the combat report; an account, it should be remembered, for which he was not responsible.

Alan Jerrard stayed in the RAF after the war, receiving his Victoria Cross at an investiture at Buckingham Palace on 5 April 1919. He was one of the RAF officers who fought in Russia in 1919, in the detachment to Murmansk. Jerrard retired from the RAF in 1933, due to ill health, and died on 14 May 1968. He was buried with full military honours at Lyme Regis on 17 May 1968.

2. Ibid.

William G. Barker's Sopwith Snipe

'Mascot' on Starboard Vickers of Barker's Snipe and Camel

William G. Barker's Sopwith Camel

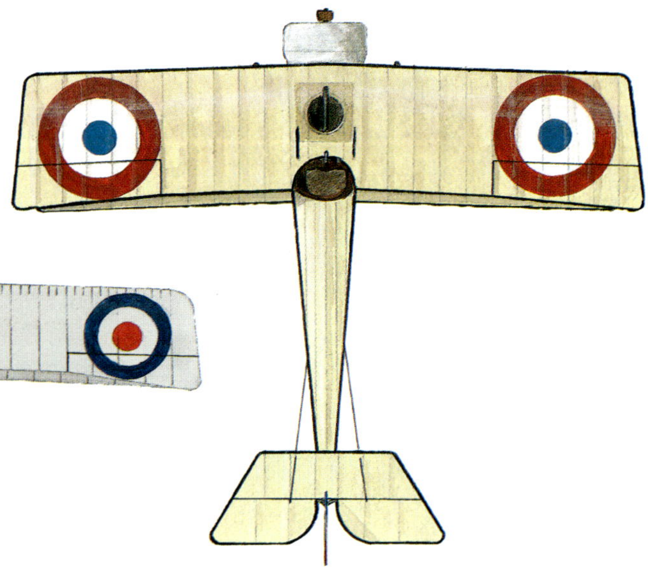

Richard Bell-Davies'
Nieuport 10

William A. Bishop's
Nieuport 17

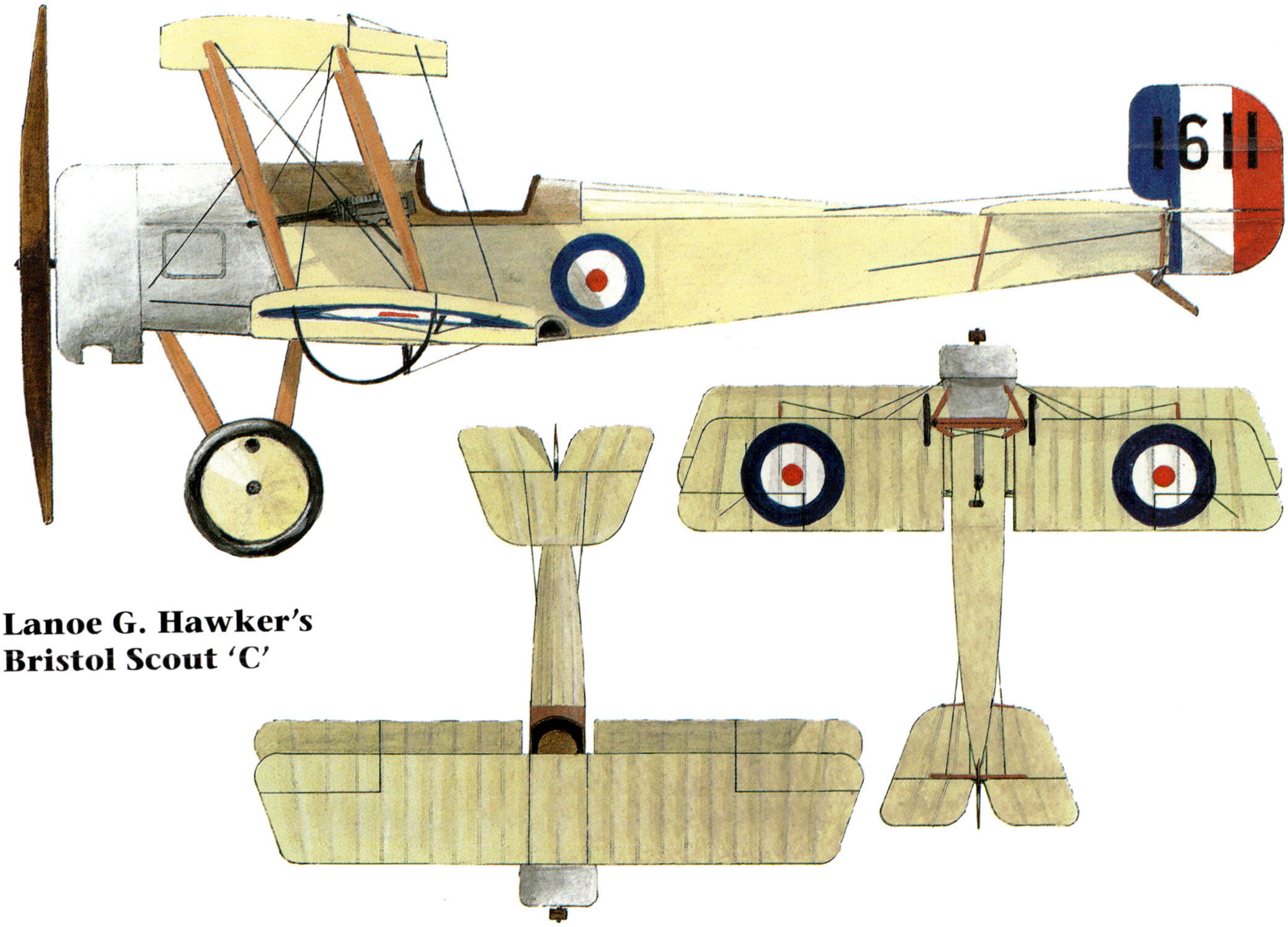

Lanoe G. Hawker's Bristol Scout 'C'

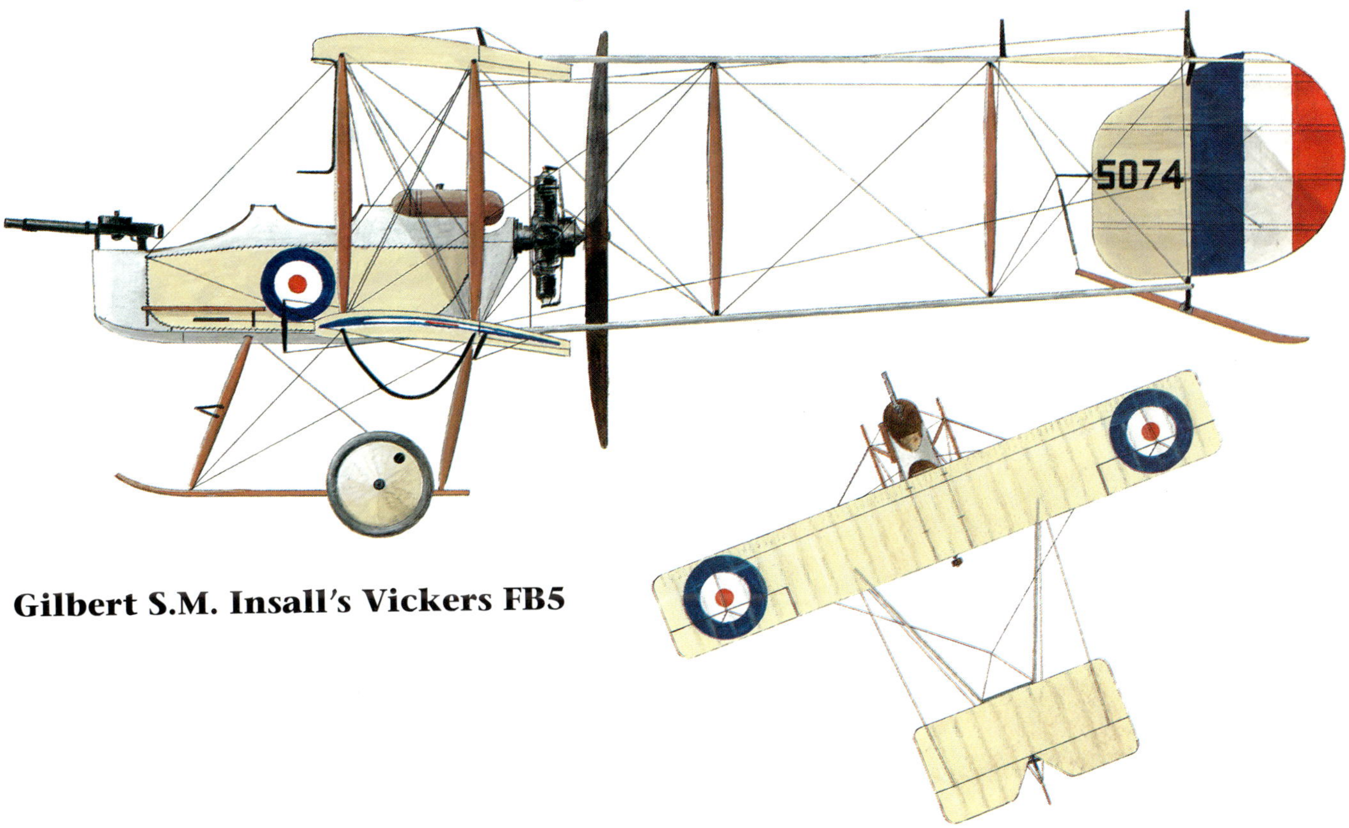

Gilbert S.M. Insall's Vickers FB5

Alan Jerrard's Sopwith Camel

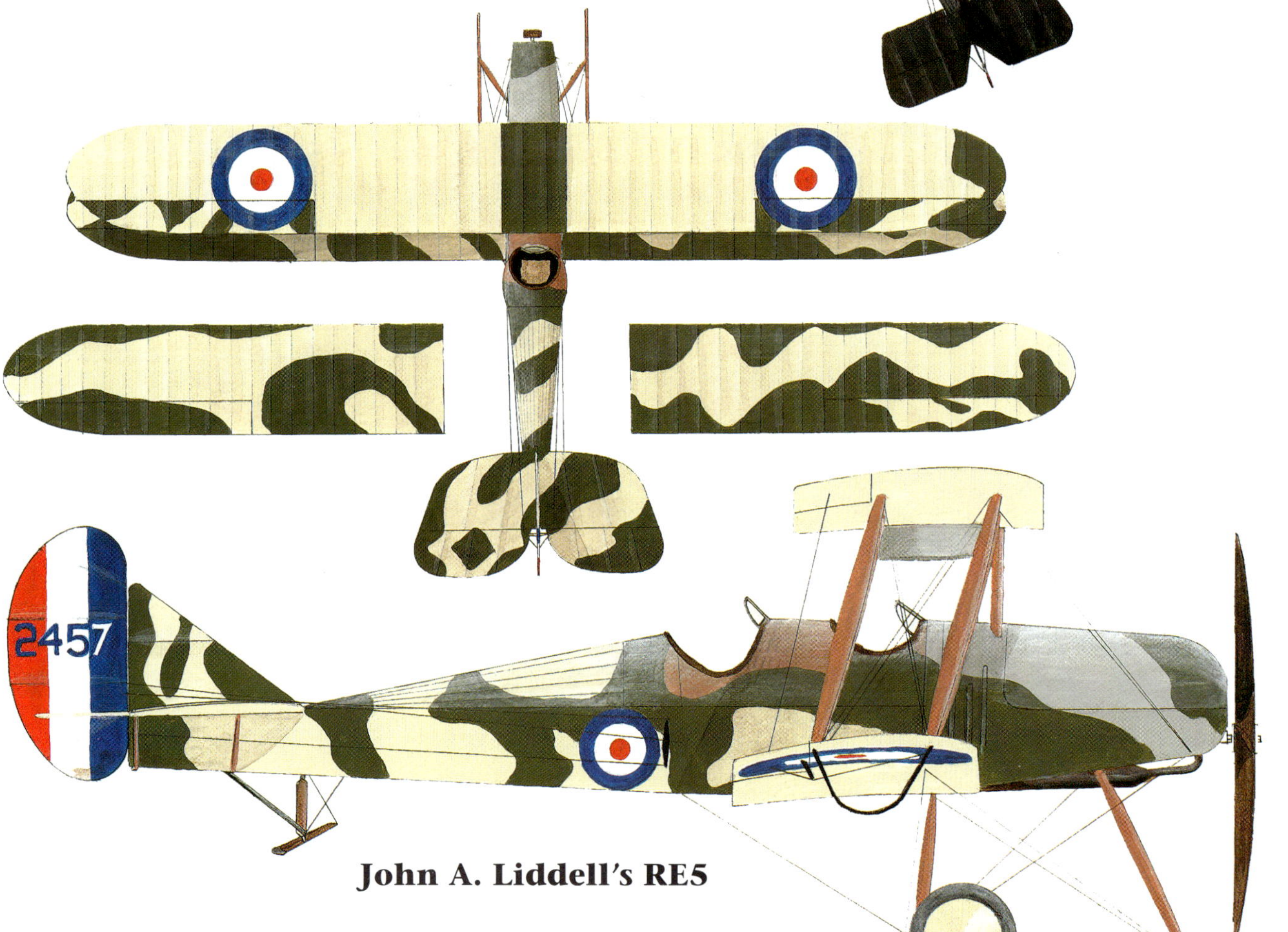

John A. Liddell's RE5

Edward Mannock's SE5a

James T.B. McCudden's SE5a

Alan McLeod's FK8

Frank H. McNamara's Martinsyde G.100

Thomas Mottershead's FE2d

Lionel W.B. Rees's DH2

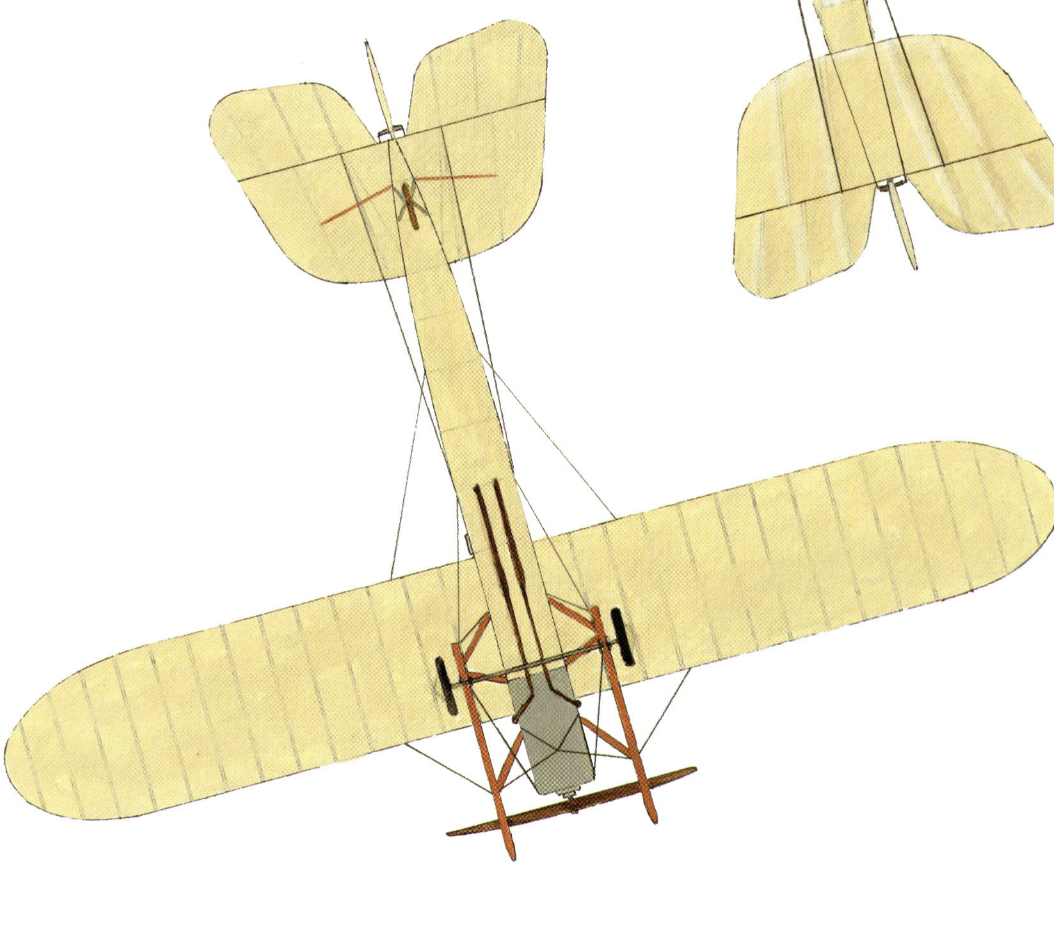

William B. Rhodes-Moorhouse's BE2b

Captain John Aidan Liddell

John Aidan Liddell was born on August 3 1888. He was a delicate boy whose early academic promise came to fruition with a First Class Honours Degree from Balliol College, Oxford. John Liddell joined the Argyll and Sutherland Highlanders on 1 June 1912, but by 1914 his enthusiasm for flying had taken hold and he learned to fly at the Vickers Flying School at Brooklands, taking his RAeC Certificate, No.791, on 14 May 1914. Liddell went to France with his regiment on 28 August 1914 as a Machine Gun Officer and he served in the trenches continuously from September 1914 until the following February, at one period spending 43 consecutive days and nights in the trenches, without benefit of a change of clothing. Liddell was invalided home to England in February 1915 and an award of a Military Cross was gazetted on 18 February. At the end of his convalescence, in May 1915, Liddell was seconded to the Royal Flying Corps on probation. After flying at Dover, Shoreham, and Farnborough, he was transferred to the RFC in July. His first posting was not slow in coming: on 23 July he was posted to 7 Squadron, based at St.Omer, joining 'A' Flight.

7 Squadron had arrived in France on 8 April 1915, with seven RE5s on charge. The RE5 was a large, commodious two-seater biplane, powered by a 120hp Austro-Daimler engine and was used for general reconnaisance and bombing. Its armament was a single Lewis gun, pistol and rifles, depending on the choice of the observer, and several 20lb Hales bombs.

Liddell flew his first patrol in France on 29 July – a routine reconnaisance of the Ostend, Bruges, Ghent area – in RE5 2458 with 2Lt. H.H. Watkins as his observer. They had several running fights with eneny aeroplanes, but returned safely. On 31 July 1915, Liddell took off again, this time flying RE5 2457, with 2Lt. R.H. Peck in the observer's cockpit. Their orders were to fly another routine reconnaissance of the same area that Liddell had flown two days before. It was late morning and the RE5 climbed steadily, arriving over Ostend at 5,000 feet. Leaving the vicinity of the town, Liddell flew on to Bruges, where he came under fire from an enemy aeroplane flying above him. Liddell was seriously wounded in the attack, a bullet hitting him in the right thigh, exposing the bone and smashing a four inch long section. With the shock and pain of his terrible wound, Liddell lost consciousness. In the front cockpit, Peck was replacing a Lewis gun drum when the RE5 dived sharply and inverted. Peck was not strapped in but he managed to keep hold of the gun mounting while everything loose in the cockpit showered past his head.

The RE5 fell for 3,000 feet before Liddell regained consciousness and managed to regain control. The German fire had been extremely accurate and damage to the RE5 was severe: the throttle controls had been shot away, as had half of the control wheel; the undercarriage was badly damaged; and the whole cockpit was badly shot about. Liddell had two possible courses of action. He could either land and get immediate attention for his appalling injuries, or attempt to return to his own lines. He scribbled a note to Peck, telling him that he was going to attempt to fly to the coast, west of Nieuport and land on the sands there, but Peck argued that the Belgium aerodrome at La Panne was a better choice. Liddell agreed. Holding the shattered control wheel in one hand and the rudder control wires in the other, and in great pain, he flew for nearly half an hour until the aerodrome at long last came in sight.

With no throttle control, Liddell was forced to land under full power, switching off the engine just before the RE5 touched down. Witnesses described the landing as 'faultless,' little suspecting that a mortally wounded pilot was at the controls. When the RE5 finally rolled to a stop it was surrounded by members of the 3rd Escadrille of the Belgian Flying Corps, eager to help, but Liddell refused to

John Aidan Liddell.

RE5 serial 2457. Liddell won his VC while flying this aeroplane.

be helped from the cockpit until a doctor arrived, fearing further damage to his leg. He applied a splint and a tourniquet to the leg himself and waited for medical aid. When a doctor finally arrived, he supervised the removal of the desperately wounded pilot from the shattered cockpit. Liddell was fully conscious and cheerful, thanking all who had helped. He was taken to La Panne hospital, where he lay for a month. He made no significant improvement, septic poisoning set in in the leg, making it necessary to amputate it in an attempt to prevent the infection spreading. The attempt was unsuccessful and John Aidan Liddell died on 31 August 1915, with his mother by his bedside. His body was taken back to England on 3 September and he was buried the next day in Basingstoke Old Cemetery, Hampshire.

Liddell's Victoria Cross had been awarded on 18 August (gazetted on 23 August) and he had been told of the award before he died. His Majesty King George V personally presented Liddell's father with his son's Victoria Cross at an investiture at Buckingham Palace.

Lieutenant Peck, whose life Liddell had undoubedly saved by his gallant and selfless act, was later killed in action on 5 March 1916 while serving with 30 Squadron in Mesopotamia.

Major Edward Mannock

In 1887, at the time of the birth of his second son, Edward Corringhame Mannock was a regular soldier, a trooper in the 2nd Dragoons, Royal Scots Guards, who had seen active service in Egypt in the campaign against Arabi Pasha in 1881–1882. In 1886, the Mannock family – Edward, Julia, and their daughter, Jessie – were living in Aldershot, where Patrick, their first son, was born, but there is some confusion over the birthplace of their second son, Edward, born on 24 May 1887. One biography gives Brighton, England, another reference gives Aldershot, but the last biography of Edward Mannock, written in 1981, and quoting family documents, gives Cork, in Ireland, where Mannock senior's regiment was stationed during the Curragh season.

When Mannock's time with the army expired, he tried several jobs, but he was unhappy in civilian life and while visiting relations in Liverpool he enlisted again and was posted to India, to the 5th Dragoon Guards. Seven months later, Julia and the three children went out to India and the family remained there for six years.

The years in India were formative for young Edward. He was a quiet, pensive boy, but curious about everything. Faced for the first time with living in an army community, he saw the social and racial injustices that were part of daily life in the British army of the time and he began to question the rightness and order of established convention which could allow such inequalities. Determined to educate himself to a higher standard than that given by army schools to the sons of NCOs, the boy became an avid reader, but one evening in 1887 he began to have difficulty with his sight and within two days he was completely blind. An examination by the army surgeon diagnosed the trouble as an amoebic infestation and under treatment the boy's sight gradually returned to normal, although it left a slight weakness in the left eye.

Mannock saw active service with his regiment during the Boer War, but on the regiment's return to England his enlistment had again expired. Once again Mannock failed to come to terms with life outside the army; he was bored with the routine of family life, his drinking, which had always been an intermittent problem, increased, and there were frequent and violent family rows. One morning, while the family were living in Canterbury, Julia found that he had gone, taking all their meager savings.

In the years following her desertion, life was extremely hard for Julia Mannock and her children. Patrick, the eldest son, was working as a clerk in Canterbury and Jessie was able to earn small sums for odd jobs, but Edward had to be taken out of school to work for a local greengrocer. The work was hard and poorly paid and the boy soon found better paid employment as a barber's assistant. In 1903, Edward met Cuthbert Gardner, a solicitor. Gardner ran the local branch of the Church Lad's Brigade, and although it was a Protestant organization, he persuaded the nominally Catholic Edward to join. This was to change Edward's life. He was thrilled by the debating club; the weekend camping trips; the meeting of people from all walks of life, and these new interests in his leisure time made it possible for him to contemplate taking an office job – a prospect he had previously detested – and he joined his brother in the offices of the National Telephone Company. But life in an office soon paled and much to his brother's disgust Edward took an outdoor job with the company – as an assistant linesman – and moved to Wellingborough, where he found a home with a Mr. and Mrs. Eyles.

It was while living with the Eyles family that Edward fully developed as a person. It was a literary and musical household, and the young Mannock made many new friends in artistic and academic circles. Despite his sketchy and largely self-won education, his natural intelligence enabled him to hold his own in the nightly discussions with the Eyles family and their friends and he gradually gained a self-confidence and maturity that he could never have reached in his own family circle. This stimulating environment helped Mannock correlate his half-formed ideas and ambitions into a cohesive philosophy.

By 1913, Mannock was full of confidence in his ability to make his way in the world; both he and Jim Eyles felt that the world was at last changing, that a new age was coming in which the mass of so-called 'ordinary' people would be able to live fuller and happier lives. But Mannock still felt stifled in England, where he felt that opportunities to better himself were still few, and he decided to try his luck abroad. He heard that the telephone company was laying a cable in Turkey and, with a loan from Mr. and Mrs. Eyles, he left England to try for a job in Constantinople. Jim Eyles later wrote: 'He was a young man with high ideals and with a great love for his fellow mortals. He hated cruelty and poverty...A kinder, more thoughtful man you could never meet...When he walked out of our little home in Mill Road, I knew instinctively that he was off to something great.'

Mannock was still working in Turkey when war was declared and when Turkey joined forces with Germany, Mannock and many of the British residents of Constantinople were interned as prisoners of war. At the beginning of hostilities the Turks had decided to deport all enemy nationals, but the Germans insisted that this decision was reversed and they were placed in camps. Conditions in the camps, for the women, children, and men, were appalling and the cruelty of the Turkish guards exacerbated the situation. Food was short and Mannock broke out of the camp several times to bring back bags of food supplied by a Turkish friend on the outside. Inevitably, he was caught and thrown into solitary confinement as a punishment. His health, already poor from lack of food, rapidly deteriorated; his body became covered in sores, which refused to heal, and he contracted dysentery. It was in this small cell, desperately ill and cut off from all human contact, that Mannock conceived his undying hatred of the German race. During the long evenings the other prisoners could hear his shouts of 'Damn the sodding Huns, bloody Huns.' Shouts which not only relieved his own frustrations and anger but were made to keep up the morale of the other prisoners.

Jim Eyles had no word of his friend for many months. In January 1915, he wrote to the American Embassy in Constantinople, asking for information. The Americans investigated Mannock's case and on 1 April 1915, Mannock was repatriated home to England on the grounds of age, bad health, and defective eyesight. It was thought that he would be of little use in England's war effort.

Mannock had been a territorial in the Royal Army Medical Corps before leaving England and on his return he applied to enlist in the Corps. His appearance was against him; he was still gaunt from the privations of his imprisonment in Turkey and looked far from well. Mannock overcame this obstacle with his usual charm, telling the medical board that he had always been thin and had never felt better; he successfully passed into the RAMC with his previous rank of sergeant.

Mannock was soon disillusioned by his essentially pacifist role in the RAMC. As his fragile health recovered, his anger at his treatment in Turkey returned; determined to get into the fighting he applied for a commission in the Royal Engineers. His intention was to become a tunneling officer, 'so I can blow the bastards up.' he wrote. Mannock's application was successful and in April 1916 he reported to the Cadet depot of the Royal Engineers at Fenny Stratford as a probationary 2nd Lieutenant.

Once again Mannock found that he was a square peg in a round hole. His fellow cadets seemed frivolous, more concerned with the best tailor they could find to style their smart uniforms, and the dash they could then cut with the local girls, than the serious business of winning the war. Some of Mannock's feeling of isolation could have been due to the fact that he was now 30 years of age, considerably older than his fellow cadets, but their mainly shallow interests, and lack of sympathy for the dangers and hardships their fellow countrymen were daily suffering in the trenches in France, infuriated and disgusted him. Always a champion of the common man, Mannock now began to feel a hatred of the British ruling class which almost equaled that which he held for the enemy.

On his next period of leave Mannock resolved to visit the Eyles, feeling sure that he would be given good advice. On the way to Wellingborough he met an old friend on the train, Eric Tomkins, who was now serving in the Royal Flying Corps. Over a drink, bringing Tomkins up to date on his last years, retelling the horrors he had seen in Turkey, Mannock confided that every move he made towards getting into the fighting in an active way seemed doomed to failure. He felt that he was merely a cog in a vast machine; what he wanted was a field of duty in which he could use his individuality.

Tomkins suggested joining the RFC, but Mannock rejected this idea, feeling that his age, general health, and rather erratic service record would be against his being accepted into yet another branch of the service. However, over the next ten days of his leave Mannock thought hard and long about Tomkins' suggestion, and he began to read all he could in the national press of the work being carried out by the RFC in France. It was now July 1916 and the whole of England was talking of the exploits of one pilot: Albert Ball. As Mannock read of Ball's successes in France, one word kept emerging from the reports – individuality. To Mannock this was the key. When he finally passed out as a 2nd Lieutenant in the Royal Engineers, he immediately applied for a transfer to the RFC.

The first hurdle Mannock had to pass to be accepted for training as a pilot in the RFC was the medical, in particular the weakness in his left eye. In addition, he knew that he would be considered old for a pilot, who were mainly in their late teens and early twenties. But Mannock was fit, and over the years had learned to compensate for the slight astigmatism in his left eye. He successfully passed the medical and was posted to the No.1 School of Military Aeronautics at Reading, to the west of London, reporting on 14 August. While at the school, Mannock was christened 'Mick' or 'Micky,' nicknames which would stay with him throughout his days in the RFC and RAF.

Completing the Reading course with honours, Mannock was posted to Hendon for elementary flying training, and he qualified for his Royal Aero Club Certificate (No.3895) on 28 November 1916. Following postings to Hounslow in December, then to the School of Gunnery at Hythe in February 1917, Mannock was posted to 10 Reserve Squadron at Joyce Green for advanced flying instruction. It was while at Joyce Green that Mannock met a pilot with similar views to his own; a pilot who was to become his great friend and mentor: James McCudden. McCudden had been posted to Joyce Green as a fighting instructor, after flying DH2s in France with 29 Squadron. He had been commissioned in January 1917 and had added an MC to his MM and Croix de Guerre, all won while he was still an NCO. In his autobiography, *Five Years in the Royal Flying Corps,* McCudden wrote of his instructing at Joyce Green: 'The pupils here...were very good. One I particularly remember named Mannock. One day he came to me and said that I was the cause of saving his life. I had only just previously given him instructions what to do if he unfortunately got into a "spin." He had just had his first spin and had remembered my advice, which I think at the time was to put all the controls central and offer up a very short and quick prayer. Mannock was a typical example of the impetuous young Irishman (Mannock was now 32 and McCudden 22. Author) and I always thought he was of the type to do or die. He now holds the DSO and MC and a Bar and at the time of writing has accounted for over two score of German machines.'

On 6 April 1917, his flying instruction over, Mannock reported to 40 squadron, based at Treizennes in France and equipped with Nieuport 17s. On his arrival at 40 Squadron, Mannock, unused to the unwritten law of the RFC in France against talking 'shop' in the Mess, created a bad impression with his fellow pilots by asking each of them how many 'Huns' they had shot down. He compounded this cardinal sin by giving them his ideas on air fighting tactics and the conduct of the war in general. The Squadron had just that day lost one of its most popular members, 2nd Lt. H.S. Pell, who had been shot down in flames by anti-aircraft fire, and Mannock expounded these views while sitting in Pell's chair. One pilot later commented that Mannock's manner, speech, and familiarity were not well received; that he seemed too cocky in view of his lack of experience. New pilots usually listened to the advice of the more experienced members of the squadron, but Mannock had his own ideas of how things should be done and offered his views on everything: how the war should be fought; their role as fighter pilots; even the suitability of their equipment. 'He seemed a boorish know-all and we all felt that the quicker he got amongst the Huns the better; that would show him how little he knew.'[1]

Mannock's first war flights with 40 Squadron did nothing to change the opinion of him formed by his fellow pilots on that first evening in the Squadron Mess. He seemed hesitant in combat situations, giving rise to unspoken criticisms of being 'windy,' there were side-long glances in the Mess and Mannock was shunned by many in the Squadron. Mannock's continual flying, practicing aerial fighting and gunnery, were seen as a show of keenness to hide his cowardice. In one such flight, Mannock had pulled the Nieuport out of its dive after shooting at the target – a white-painted circle on the edge of the aerodrome – when the starboard lower wing twisted out of its retaining socket, a not uncommon fault on the Nieuport 17. Mannock fought to regain control and managed to force land in a field close by the aerodrome. Mannock was unhurt and his fellow pilots had to admit that he had shown courage, along with flying skill, in successfully landing the Nieuport.

On the morning of 7 May, Captain W.E. Nixon led five Nieuports to attack a line of observation balloons five miles behind the enemy lines. The Nieuports – Lts. Scudamore, Hall, Redler, Parry, and Mannock – flew at extremely low altitude, a plan of attack devised by their commanding officer, Major Tilney, and which had been tried successfully a few days before. Flying through intense fire from the ground defenses of the balloons, the Nieuports began to attack their individual targets. But Nixon had seen a flight of Albatroses above the balloon line. He turned away from his balloon and climbed to meet the threat, hoping to keep the enemy scouts from attacking the Nieuports. Nixon was hopelessly outnumbered and was almost immediately shot down by Lothar von Richthofen of *Jasta* 11.

All the other Nieuports were severely damaged. Hall and Scudamore crashed on landing back at 40 Squadron's base at Bruay; Parry's Nieuport was shot in the fuel tank and he crash-landed just behind the British lines at Camblain L'Abbé; Redler force landed at Savy but

1. *Mick. The Story of Major Edward Mannock*. James M. Dudgeon. Robert Hale, London. 1981.

Edward Mannock.

damaged his Nieuport on returning to Bruay. That night, Mannock wrote in his diary:

'All except the Captain returned safely with machines almost shot to pieces...I was the only one to return properly to the aerodrome and made a perfect landing. We all got our objectives. My fuselage had bullet holes in it, one very near my head, and the wings were more or less riddled. I don't want to go through such an experience again.'[2]

Two days later Mannock flew a patrol with Captain Keen leading. One of the Nieuports dropped out with engine trouble but Keen and Mannock carried on and attacked an enemy two-seater over Henin-Liétard. The Nieuports were then attacked in turn by three Albatroses. Mannock was separated from Keen and his Lewis gun jammed, the Aldis sight oiled up, and the engine of the Nieuport cut out at 'the crucial moment.' 'I thought all was up.' Mannock later wrote. He threw the Nieuport into a spin, pulled out lower down and zigzagged for the safety of the British Lines. Mannock's engine picked up again at 3,000 feet over Arras and he climbed back to 12,000 feet, where he met another enemy machine. 'I hadn't the pluck to face him. I turned away and landed here with my knees shaking and my nerves all torn to bits. I feel a bit better now but all my courage seems to have evaporated after that experience this morning.'

It is obvious that at this time Mannock had still not conquered his nerves and the other pilots in the Squadron still considered him as 'windy.' Over the next few days he tried desperately to come to terms with his fears. Mannock was a sensitive and complex man and was intelligent enough to realize the dangerous nature of the work he was doing. He was also nearly a decade older than many of his fellow pilots, without that sureness of immortality that youth often gives.

On 12 May an enemy machine flew over Bruay. Mannock was the first off the ground in a futile attempt to intercept the enemy and his obvious keenness brought about a slight thaw in his relationships with the other pilots. People began to speak to him in the Mess, buy him drinks, and he was included in the trips by the Squadron tender into the nearest town to let off steam. Major Tilney also did his best to help, on occasions giving Mannock the responsibility of leading a Flight.

On 25 May, Mannock attacked a two-seater, firing 30

2. Ibid.

rounds of Lewis into it around the area of the pilot's seat. The two-seater dived away, but although Mannock felt sure he had sent it down he did not claim it. None of his companions had seen the fight and he did not wish to claim an unconfirmed victory, fearing their comments. That he felt sure of his success was enough. Just over a week later Mannock had another combat with a two-seater, emptying a drum of Lewis into it from 25 yards range. The enemy pilot put down the nose of his machine and dived away. Mannock again made no claim, but his new aggressiveness had been noted and the thaw in relations continued.

On 7 June, Mannock scored his first confirmed victory, an Albatros D.III shot down out of control at 13,000 feet north of Lille. The Albatros was confirmed by the crew of one of the FE2ds of 25 Squadron that Mannock was escorting and the victory was shared with them.

A week later, coming in to land, Mannock suddenly felt a searing pain in his right eye and his vision was completely obscured, both his eyes streaming water. Mannock somehow managed to land the Nieuport and was taken to the first aid tent, but attempts to treat the eye caused him to faint. He was taken to the nearest hospital, cocaine was injected into the badly inflamed eye to deaden the pain and the surgeon removed a large piece of grit and a piece of metal, which had come from the cowling of the Nieuport. Although Mannock was fully recovered from the accident in three days, he was sent home on leave.

Mannock spent an unhappy few days with his mother. She felt abandoned by her children, was drinking heavily, and insisted that Mannock send her more money. Mannock realized that he could no longer be a part of such a broken and unhappy family – he had changed too much – and he spent the rest of his leave with the Eyles. While with Jim Eyles he discussed the problems he had been having: his attempts to conquer his fears and his cautious approach to combat while he refined his flying and gunnery skills, an approach not understood by his fellow pilots. He confided that he had realized that his age was against him in this new form of fighting, but he could compensate for this by superior tactics and careful planning. He now felt he was ready.

Mannock returned to 40 Squadron on 2 July; ten days later he shot down a DFW for his third victory. Sighting two DFWs south east of Lens, he climbed for height and attacked the rearmost enemy machine from behind, firing 90 rounds from his Lewis gun. The DFW fell away out of control and Mannock watched it turn over onto its back and crash behind the British lines at Avion. Here, at last, was a positive victory. Landing at Bruay, Mannock went by tender to the site of the wrecked DFW. The observer had survived the crash and had been taken away, but the dead pilot lay nearby, covered with a gas cape. Mannock lifted the cape, forcing himself to view the results of his work. He then moved over to the remains of the DFW. In the observer's seat was a little black terrier, dead. 'I felt exactly like a murderer,' Mannock wrote.

The next day, Mannock shot down another DFW, this

The Mannock family. Edward at left with his mother, brother Patrick, and sisters Jessie at the rear and Nora in front.

one out of control over Sallaumines. An hour later he attacked a DFW near Douai, but held his fire too long and overshot, having to bank away to avoid a collision. Turning back, Mannock fired 30 or 40 rounds into the enemy machine and it dived away towards Douai.

The new Mannock was now mainly accepted by his fellow pilots; only two refused to recognize the change in him since his return from leave and his recent successes, still being outspoken in their comments on what they saw as his lack of. Factions arose and a rift developed between the Mannock supporters and Godfrey and Hall, his chief detractors. Tilney became concerned by the effect of this on the morale of his squadron and he asked Mannock's Flight Commander, Captain George 'Zulu' Lloyd, to have a heart to heart talk with Mannock. Lloyd later wrote: 'He was not actually called yellow, but many secret murmurings of an unsavoury nature reached my ears. He showed signs of being over careful during engagements. He was further accused of being continually in the air practicing aerial gunnery as a pretense of keenness – in other words, the innuendo was that he was suffering from cold feet.'

Lloyd talked to Mannock, telling him of these accusations. Mannock admitted to having been very afraid, much against his will, but he had now conquered his fears. He explained that his cautiousness was due to his wishing to learn the tactics of airfighting. He had no wish to be killed before he had had a chance to learn his trade, as were so many young pilots. He now felt that he had conquered his fears and had evolved his tactics. Lloyd accepted Mannock's honest and straightforward explanation.

While Mannock was attacking an enemy machine on the morning of 14 July, a tappet rod broke, ripping off the cowling of the Nieuport, Gliding towards the British lines from 15,000 feet, Mannock put the Nieuport down in a field of tall corn, turning it over onto its back. Mannock was unhurt. That morning 40 Squadron lost another pilot: Lt. G. Davis was shot down by *Ltn.* Erwin Böhme of *Jasta* 29 and taken prisoner.

On 20 July, Captain Keen led a patrol of six Nieuports across the lines, flying north to Armentières. The Flight was attacked by 'three of the finest Hun pilots I ever wish to meet.' Mannock later commented in a letter home. The fight lasted ten minutes, but neither side could score any advantage and the German pilots, outnumbered two to one by the Nieuports, eventually broke off the combat. Mannock wrote: 'I shall always maintain an unsullied admiration for those Huns. The aircraft battery people reported the battle as one of the most splendid exhibitions of tactics they had ever seen. We did nothing but swear. Likewise the CO.'

On 21 July, 40 Squadron suffered a loss which deeply affected the pilots. Second Lieutenant F.W. Rook, only 19 years of age and one of the most popular members of the Squadron, was shot down in flames over the Cambrai to Douai road by *Oberleutnant* Tutschek of *Jasta* 12. The next day, before the pilots had had time to come to terms with Rook's death, another popular member of the Squadron was killed, a direct hit from 'Archie' blowing off the wing of William Bond's Nieuport, which broke up in mid-air.

On 22 July, the day of Bond's death, Mannock was awarded the Military Cross, promoted to Captain, and given command of a Flight. The citation spoke of 'many combats' and 'very fine offensive spirit and great fearlessness attacking the enemy at close range and low altitudes under heavy fire from the ground.' Trenchard sent his personal congratulations. The wording of the citation and the telegram from Trenchard seem at odds with the reports of Mannock's reputation in the Squadron, yet the recognition of his work seemed to reopen the recent doubts. Mannock's promotion to Flight Commander was to give fresh impetus to his detractors, who, while admitting that he had had some success as a lone fighter, doubted his ability to lead a Flight. Their misgivings seemed justified when Mannock returned from his first patrol as a Flight Commander. He had handled the patrol badly, several times placing it in danger of attack under unfavourable circumstances, so much so that his great friend in the Squadron, Lt. William MacLanachan – nicknamed 'McScotch' by Mannock – had left the formation in disgust at Mannock's leadership.

On 5 August Mannock shot down an Albatros out of control over Avoin for his fifth victory. A week later the Squadron received a call to say that an enemy fighter was attacking an observation balloon. Mannock took off from the advanced landing ground at Mazingarbe to intercept the intruder, finally finding the Albatros as it was about to attack another balloon. The Albatros made for the safety of its own lines, but Mannock followed, firing short bursts to force the enemy pilot to take evasive action and lose speed. As Mannock closed the range the Albatros flew straight and level for a few seconds and Mannock fired a burst of 20 rounds, wounding the pilot, *Ltn.* Joachim von Bertrab of *Jasta* 30 in both arms and one leg. Bertrab managed to land the Albatros and was taken prisoner. Mannock was delighted: 'two machine guns with a thousand rounds of ammunition against my single Lewis and three hundred rounds,' he exulted, adding, 'he didn't hit me once.'

Mannock had been deeply hurt by what he saw as MacLanachan's desertion during the patrol of 23 July, but it was not until he repeated his mistake while leading another patrol that the lesson was driven home. During a dinner to celebrate his downing of von Bertrab, a plan was conceived to attack the enemy aerodrome at Dorignies and Mannock led Lts. Kennedy and Tudhope off the ground at 6:00a.m. on 13 August. Before the three Nieuports had reached their target they sighted a formation of nine enemy scouts. Undeterred by the odds of three to one, Mannock instantly attacked the Albatroses. But the German pilots were all experienced and fought the three Nieuports with skill and great determination. It was only by luck that the British pilots were finally able to extricate themselves from the situation and return to Bruay. All the Nieuports were badly shot about, but Tudhope's was the worst hit: a main wing spar had been cut through, the fabric of the top wing was in ribbons, all the instruments shattered, and a bullet had gone through the collar of Tudhope's flying coat. Mannock was severely shaken by the realization that his bad leadership had nearly caused the death of his friends and himself. The lesson clear: he would have to drastically rethink his tactics if he was to become an efficient patrol leader. This patrol marked another stage in Mannock's development as an airfighter. He was to become one of the RFC and RAF's finest patrol leaders of the war.

This further change in Mannock was soon only too apparent to the members of his Flight. He no longer went bull-headed at the enemy, but carefully stalked them, placing the Flight in an advantageous position before attacking. Over the next weeks Mannock daily honed these new skills.

On 15 August Mannock scored a 'double,' shooting down an Albatros just after mid-day over Lens and destroying another, probably from *Jasta* 30, north of the town during an evening patrol. Two days later a DFW fell under his guns, north east of Sallaumines, bringing his victory total to nine.

Mannock was again brought face to face with the grim reality of war on 22 August when his friend Lt. H.A. Kennedy was killed during a fight with seven enemy fighters near Douai. Mannock did his best to rescue Kennedy, but was too late. He watched helplessly as the

Mannock in the cockpit of his SE5a at London Colney.

silver Nieuport with the large white number '4' on its top plane, caught by the light of the evening sun, dived ever more steeply. That night MacLanachan heard Mannock crying in his hut and entering saw that Mannock was sitting on the edge of his bed, rocking backwards and forwards – 'keening' in the ancient Irish way for his dead friend.

Like his friend McCudden, Mannock was now becoming expert at destroying enemy two-seaters, shooting down a DFW on 17 August, another four in September, plus an unidentified 'C' type and a Rumpler.

The DFW destroyed on 4 September had a traumatic effect on Mannock. He first saw the enemy two-seater flying south west at 10,000 feet over Avoin. Mannock was below the DFW and could not make out any national markings, so he turned and flew under the two-seater reasoning that if were British it would take no notice of the Nieuport, but if hostile it would attack. The ruse worked. The enemy pilot, seeing what he thought was an easy target, put his nose down and fired a burst at the Nieuport. Mannock immediately turned, dived for speed, then zoomed and turned again, taking up a position behind the enemy machine. Closing the range and ignoring the return fire from the enemy gunner, Mannock fired 50 rounds in short bursts. The DFW burst into flames, shedding pieces of wings and tail as it fell. What had moments before been a graceful yet powerful aeroplane of wood and metal, flown by two young men like himself, was now an unrecognizable blazing mass. Mannock was sickened at the sight. He had seen burning machines before, but never one which had been destroyed by his own hand. The memory of this burning DFW always stayed in his mind and it was perhaps this combat which convinced him that he too would die in a burning aeroplane. The DFW from *Flieger Abteilung* 235 was crewed by *Uffz.* Georg Frischkorn and *Ltn.d.R.* Fritz Frech. Several months later a note was dropped behind the British lines, addressed to 'The British Flying Corps.' It read:

'The 4th September I lost my friend Fritz Frech. He fell between Vimy and Lievin. His respectable and unlucky parents beg you to give any news of his fate. Is he dead? At what place found he his last rest? Please to throw several letters, that we may find one. Thanking before. His friend K.L.

PS. If it is possible, send a letter to the parents: Mr. Frech, Kónigsberg 1. Pr.Vord Vorstadt 48/32.'[3]

Mannock was awarded a Bar to his MC. on 14 September. He was now fully confident in his role as Flight Commander. Now that he had proved to himself that his tactics were correct, and that he had largely overcome his fears, he was no longer interested in his own prowess and considered that he was merely doing his 'work,' an important part of which was looking after the members of his Flight – especially the 'new boys.'

In late August a replacement pilot had been posted to 40 Squadron: Lt. George McElroy. McElroy was to become a renowned airfighter, but his start in 40 Squadron – like Mannock's – was not auspicious: he crashed two Nieuports in one day. Tilney was in favour of sending McElroy home for further instruction, but Mannock successfully argued for McElroy's retention and took the new boy under his wing, naming him 'McIrish' to differentiate him from 'McScotch' – MacLanachan.

Mannock scored no further victories in the remaining months of 1917. At the beginning of October the weather became bad with high winds and rain. The conditions were welcomed by the tired pilots. Major Tilney wrote home. 'Thank goodness the weather has broken. I hope it will be dud for a bit. My fellows are dead beat. Six weeks solid flying.'

On 17 October the new SE5as began to arrive. The SE5a was powered by a V-8, water-cooled stationary engine and the 40 Squadron pilots, used to the rotary-engined Nieuports, began to convert onto the new type. Few patrols were flown and only two squadron victories were claimed during the month: an Albatros to Captain G.L. Lloyd on 7 October – the last victory claimed on a Nieuport – and a two-seater to MacLanachan on the last day of the month for the Squadron's first victory in the SE5a.

At first the SE5as were viewed with optimism: it was faster than the Nieuport and, more importantly, armed with two machine guns, but the Hispano-Suiza engine of the SE5 was unreliable and new to the 40 Squadron mechanics. There were more than 20 forced landings in two weeks attributable to the faulty engines. What patrols were flown in the new machines were almost

3. Mannock later received this letter while on leave in England and wrote to the parents of Fritz Frech.

invariably depleted by engine failures. In three weeks the pilots became disillusioned with both the engines and the guns – which continually jammed – of the SE5a and morale suffered.

By the end of December 1917 Mannock had been ten months with 40 Squadron. It was obvious to his friends, especially to MacLanachan, that he was tired and in need of a rest. On the first day of the New Year 1918, Mannock scored his 16th victory, a DFW shot down into the British trenches at Famboux. The next day Mannock left 40 Squadron, posted to Home Establishment. On Mannock's departure, Major Tilney wrote in the 40 Squadron diary: 'His leadership and general ability will never be forgotten by those who had the good fortune to serve under him.'

Before his leave had ended Mannock was impatient to return to France and the life of a squadron on active service, but when he next reported for duty he was posted to The Wireless Experimental Establishment at Biggin Hill, where he flew FE2s on radio testing flights. It was a boring and, to Mannock, an unrewarding job, but he had only been at Biggin Hill for a short while before he was posted as a Flight Commander to 74 Squadron, working up to strength at London Colney in Hertfordshire.

Mannock arrived at London Colney on 2 February 1918. The nucleus of the new squadron had been at the aerodrome since the previous July, flying several types of aircraft, and a combination of bad weather and disinterested and aloof instructors had resulted in low morale. Several of the trainee pilots had seen action in France as observers and were unimpressed with the quality of their instructors and the extra privileges they enjoyed. The arrival of Mannock changed this. Here was a pilot who was only too anxious to give them the benefit of his experience, somebody to whom they could relate. His lectures on airfighting were fully attended and were full of interest and advice. Time and time again he expounded his dictum: 'Gentlemen,' he would start each lecture, 'always above; seldom on the same level; never underneath.'

On 7 March, the trainee pilots were anxiously waiting to see which of them would be chosen to fly to France with 74 Squadron. Thirty two trainees and seven instructors were in competition for the 20 places. When the results were finally announced the three Flight Commanders were: Captain W.E. Young, 'B' Flight; Captain W.J. Cairns, 'C' Flight; and Captain E. Mannock, 'A' Flight. The Flights were then sent to Ayr, in Scotland, to the No.1 School of Aerial Fighting, for a course on aerial fighting and tactics. On their return to London Colney they found that they had a new commanding officer: Major Keith Logan Caldwell, a large, black-haired New Zealander. 'Grid' – as Caldwell was known throughout the RFC – had served continuously in France from July 1916 to September 1917, with only one break of seven weeks on sick leave. He had scored one victory with 8 Squadron, flying BE2ds, before being transferred to 60 Squadron in November 1916. During his seven months with 60 Squadron, flying Nieuports, Caldwell had scored another eight victories, been awarded an MC, and promoted to Flight Commander. The new CO was a man after Mannock's own heart and they became great friends.

Over the following days 74 Squadron received their complement of 19 SE5as, and on 25 March flew them to Goldhanger aerodrome in Essex. On 30 March, the Squadron left for St.Omer in France. After staying for a few days at several aerodromes, 74 Squadron finally settled into its base at Clairmarais North on 11 April 1918. Mannock was back at war.

On 12 April, 74 Squadron flew its first patrol over the Lines, 'C' Flight taking off at 6:00a.m. The Flight had a fight with a formation of enemy scouts but scored no victories. Mannock led 'A' Flight in the next patrol of the day, leaving at 8:25a.m. Twenty-five minutes after leaving the ground, the SE5as ran into a group of Albatroses over Merville at 13,000 feet and Mannock shot down one and Dolan claimed another. Mannock led another offensive patrol later in the day and it attacked a yellow and black Albatros, shooting it down near Bois de Phalempin. Although all the members of the Flight shot at the Albatros, the pilots were certain that it was Mannock's fire which had sent it down, but in his report Mannock insisted that it should be shared by the whole Flight. This was the first of many instances where Mannock disclaimed a victory as his own, giving it to the Flight as a whole, or to a new pilot for his first victory, as a boost to morale. A number of pilots who flew with Mannock in 74 and 85 Squadrons have independently commented on this habit of Mannock's. It seems that at this stage in his career he simply was not interested in his personal score. The safety and efficiency of his Flight was of paramount importance and nothing else mattered. This giving of victories to others in order to instill confidence was not unique to Mannock. Caldwell later commented. 'It has been said often that Mannock gave credit to one of his juniors for shooting down a Hun that he, himself, had actually hit. I believe this to be true, and actually know of others who have done the same. The team spirit was very strong in the better squadrons. In entering a scrap it was usually the leader who fired the first shots, but often his followers would fire on the same aircraft (sometimes needlessly) and this would be a chance for a leader to tell "so and so," on returning to the ground, that he got the Hun. His logbook, however, might not indicate this generous action.'

The Welshman, James Ira Jones, was a pilot in 74 Squadron at this time, and he was later to write a personal history of 74 Squadron during its days in France in 1918, and the part Mannock played in its success. Mannock served in 74 Squadron for only three months and five days before he was promoted to Major and given command of 85 Squadron, but he was now a superb airfighter and in that short period he scored 33 confirmed victories, credited to him alone, and shared in three more. But it is obvious from the recollections of Caldwell and others that he actually destroyed many more while with the Squadron.

Jones – known to all as 'Taffy' – who was himself to become a renowned airfighter with 37 victories by the end of the war, has given us a vivid picture of Mannock in combat during his days with 74 Squadron. On 21 May Mannock scored four times, shooting down an Hannover two-seater during a morning patrol and three Pfalz D.IIIs in the evening. Ira Jones wrote:

'In his (Mannock's) first fight, which commenced at 12,000 feet, there were six Pfalz scouts flying east from the direction of Kemmel Hill. He shot one to pieces after firing a long burst from directly behind and above; another he crashed. It spun into the ground after it had been hit by a deflection shot. Then Mick had a fine set-to with a silver bird while his patrol looked on. It was a wonderful sight. First, they waltzed around, with Mick tight on the bright lad's tail. Then the Pfalz half-rolled, falling a few hundred feet beneath him. Mick did the same, firing as soon as he got his enemy in line. The Hun looped. Mick looped too, coming out behind and above the other, firing short bursts. The Pfalz spun. Mick spun also, firing as he did so. This shooting seemed to me a waste of ammunition. The Hun finally pulled out. Mick, who was now down to 4,000 feet, did the same. The Pfalz started twisting and turning, a sure sign of "wind-up," and Mick administered the coup de grâce with a burst from directly behind at about 25 yards range. The Hun went down, obviously out of control, and crashed.

This really was a remarkable exhibition of cruel, calculated Hun-strafing. I felt sorry for the Hun. He put up as fine a show of defensive fighting as I've ever seen...I asked Mick after we had landed why he had fired during the spin. He replied: "To intensify his wind-up."' [4]

During the afternoon of 22 May, 74 Squadron was visited by General Plumer, commanding the Second Army. He asked which of the assembled pilots was Mannock. Mannock was pointed out. The pilots of 74 Squadron were not noted for their sartorial elegance, but Mannock, who had just landed from a patrol, was the most disreputable figure of them all. He was hatless, his tunic was open, he was not wearing a collar and his hair was uncombed. If Plumer was shocked by this apparition, he pulled himself together and extended his hand to Mannock. Mannock gave Plumer his usual hearty handshake; the General was seen to wince slightly, then he stammered: 'Mannock, let me congratulate you on your DSO.'

By the end of May, eight of Mannock's victories while with 74 Squadron had fallen in flames and it was about this time that his old conviction, that he would finally meet his own end in a like fashion, again surfaced. Jones has related how Mannock was getting 'very peculiar over the business,' telling how Mannock would come into the Mess after scoring a victory in flames, 'whooping and halloing; "Flamerinoes, boys! Sizzle, sizzle, wonk."' After describing with great relish the feelings of the German pilot, Mannock would then turn to one of the company and say 'That's what will happen to you on the next patrol, my lad.' But later Mannock would be seen in a gloomy mood, elbows resting on his knees, his chin in his hands. Asked what was the matter, he would reply. 'That damned Hun in flames.' Jones realized that Mannock was becoming obsessed with the idea of being shot down in flames. Occasionally Mannock would ask why they had not been issued with parachutes, as were the German pilots.

That Mannock's nerves were now in a bad way is evidenced by the growing intensity of his hatred for the enemy, who he feared might win the war. After shooting down an enemy two-seater, which managed to land safely, Mannock dived half a dozen times, firing at the pilot and observer. Caldwell was flying with Mannock on this occasion and flew alongside him, 'yelling at the top of my voice (which was rather useless) and warning him to stop.' Caldwell took Mannock to task after they had landed but Mannock replied, in temper, 'The swines are better dead – no prisoners for me.'

This behaviour of Mannock's is at variance with all who knew him as a man of warm, generous, and full-hearted nature, kindness personified. Caldwell later wrote that he did not consider that Mannock's nerves were strained, almost to breaking point, when Mannock left 74 Squadron to take command of 85 Squadron, but Mannock was adept at hiding his true feelings. When Captain Cairns was shot down and killed on 1 June, the pilots held a party that night in his honour. During the party Mannock seemed his usual cheerful self, but in the early hours of the morning, as the party was dying, he took up his violin and began to play. His deep sorrow was evident from his facial expression and in every note he played. Told of this by the pilots, he would only continue playing while facing the wall. In his hut later, Jones recalled, Mannock sat on the edge of his bed, arms wrapped around his body, desperately trying to stop his trembling.

Mannock was awarded a Bar to his DSO on 8 June; on 18 June he left France for a fortnight's leave. He spent this leave with the Eyles family and it was only too obvious to Jim Eyles that his friend's nerves were extremely bad. The old wit and sparkle were gone; his hands continually shook and he would suddenly begin to tremble with a frightening violence. At other times he would cry uncontrollably, muttering something which Eyles could not understand. His face would become wet with tears and saliva, soaking the front of his shirt. He tried to make light of his condition to Eyles, obviously feeling ashamed of what he saw as a weakness, but he could not stop. 'I feel better for a good cry,' he said. When Mannock left at the end of his leave, Jim Eyles sensed the difference from other good-byes. He knew that Mick Mannock was a sensitive man, who was somehow sure that he would not be returning from France again; that he was saying a final good-bye.

Mannock took command of 85 Squadron on 5 July 1918. The Squadron was not highly motivated and its morale was low. Major William Bishop VC, its first commanding officer, had been a lone-wolf type of leader and had left the rest of his pilots to fend for themselves. The Squadron was badly in need of someone to lead and shape them into a team. Even sick, as he undoubtedly was, Mannock was just the man for the task. He realized that he had excellent pilots in his Squadron – men of the caliber of Captain Spenser Horn and the Americans, Larry Callahan and Elliot White Springs – but they all needed to be instilled with a team spirit, which he had taught so

4. *Tiger Squadron.* Ira Jones. W H Allen. London 1954.

Captain Edward Mannock while serving in 74 Squadron in the early summer of 1918.

well in 74 Squadron, and given that sense of direction and purpose.

As a squadron commander Mannock had no need to fly patrols – in fact the flying of patrols by squadron commanders was actively discouraged by Wing – but that was not Mannock's way; he led by example, and after only two days he had literally transformed the morale of the entire squadron by his teaching of tactics, both in theory on the ground and in actual combat. Mannock was inwardly still tense, but the violent mood swings seemed to have gone. The training of his pilots gave him a more settled outlook and he became more at ease with himself. However, this new sense of relative well-being was dealt a severe blow on 10 July, when he heard that his friend James McCudden had been killed in a flying accident the previous day. If the great McCudden could be killed, who else could possibly survive. Mannock's hatred of the enemy returned in full force. Since taking up his command he had only claimed two personal victories, but in the two weeks after the news of McCudden's death he flew many patrols, both leading the Squadron and alone, shooting down six enemy machines. The swings of mood were back in full force and Mannock talked to his close friends that he was certain he was moving towards 'something final.'

Ira Jones wrote in his diary for 25 July: 'Just returned from 85. Had lunch, tea and dinner with Mick. I can't quite make out whether he has got nerves or not. One minute, he's full out. The next he gives the impression of being morbid and keeps bringing up his pet subject of being shot down in flames.' On Jones telling Mannock that he had shot down a two-seater in flames that morning, Mannock replied with a sour smile. "Could you hear the sod scream. One day you'll get it like that, my lad. You're getting careless. Don't forget to blow your brains out." 'Everyone roared with laughter.' But it was evident that Mannock's teaching had transformed the spirit in 85 Squadron. In the same entry, Jones commented: 'The squadron is very full out. Everyone says they much prefer Mick to Bishop as a CO. Mick leads all the squadron patrols and produces the Hun on a plate. Bishop rarely did this.'

Before Ira Jones left to return to 74 Squadron, Mannock took him into his hut. He wanted to talk. Jones asked how he felt. Mannock's reply shocked him. He felt that he would not last much longer, he said. With tears in his eyes he told Jones that it did not matter if he were killed, he felt that he had done his duty to the best of his ability and he would be in good company. Mannock then cautioned Jones that he had been told that he had been reckless on more than one occasion; he was becoming dangerously over-confident, Mannock warned. Jones wrote. 'He is a grand fellow. I love him.' Then Mannock's mood changed and linking arms with Jones he walked back into the Mess singing 'Rule Britannia.' It would not do for his pilots to see their commanding officer in a solemn mood.

In the early morning of 26 July, Mannock and a new pilot, Lt. Donald Inglis, took off and made for the lines. Inglis had not yet been successful in scoring a victory and Mannock had promised the previous day to take him out and show him how it was done. Mannock had given Inglis full instructions. If they found a two-seater, then he, Mannock, would attack first. Then Inglis would come in close behind his leader and underneath. When the enemy machine was full in his sights he would simply fire a long burst into it. It would go down.

At 5:30a.m., at 5,000 feet over Merville, the SEs spotted a two-seater, low down over the lines. Mannock dived, with Inglis following, and attacked, shooting the observer and leaving Inglis with a defenseless target. Inglis, following his instructions to the letter, came up underneath the enemy machine and fired, hitting the petrol tank and almost ramming the two-seater before pulling away. The two-seater went down in flames and crashed at La Cix Marmuse. The two SEs were now very low over the German side of the lines, but Mannock then flew twice around the burning wreck, a most unusual thing for him to do and against all his teaching, before finally turning for the British lines. Inglis, following, saw Mannock kick his rudder and realized for the first time how low they were. Suddenly a small flame came out of the side of Mannock's SE5a and it grew larger as the horrified Inglis watched. Mannock was no longer using his rudder and the nose of the SE dropped, turned twice in a slow right hand turn and hit the ground in flames. Inglis circled the spot at 20 feet but was hit by groundfire in the petrol tank and he made a forced landing a few yards behind the British front-line trenches. Inglis was helped out of the damaged SE5a by Welsh troops who were holding the line and sat down on a firestep. His nerves were shattered. All he could say was: 'The sods have killed my Major.'

The pilots of 85 Squadron were at breakfast when the telephone rang. Horn answered it. An anti-aircraft battery near Hazebrouck was on the line. Had Major Mannock returned. Horn asked why. The answer was that a couple of SEs, one with a red streamer on its tail, had shot down a German two-seater at 5:30a.m. but had not been seen to return. 'We are sure it was Major Mannock, because we know his tactics.' The squadron thought this was so much rot. How could anyone recognize Mick's flying, and plenty of leaders flew a red streamer. Besides, who could shoot down Mick. He was invincible. Half an hour later the Squadron's telephone rang again. The orderly room clerk took down the message in a trembling hand. 'Major Mannock down by machine gun fire from ground between Calonne and Lestrem after bringing E.A. two-seater down in flames at Lestrem. Lieutenant Inglis shot through petrol tank and landed on front line at St.Floris. Machine O.K., Pilot O.K. Machine likely to be shelled, salvage tonight if possible, more later. Machine at Sheet 36a S.E. or 36 N.W.K.31,D.14.'

The pilots of 85 and 74 Squadrons were stunned by the news of Mannock's death. 'Grid' Caldwell and Ira Jones flew over the area the next morning, but there was nothing to see and the intense groundfire forced them to abandon their search.

Over the years researchers have put forward several theories as to why Mannock followed the two-seater down so low and circled the crash site, pointing out that it was against all his teaching and completely out of character, but many very experienced fighter pilots – in both World Wars – have departed from their own teaching and experience for one instant and paid the price with their lives – Manfred von Richthofen being a prime example. One theory advanced was that Mannock wanted to die, that he deliberately put himself in a position on the morning of 26 July which made it almost certain that he would be shot down. This is an untenable argument. Mannock was a complex man, but one aspect of his character is crystal clear: his strong sense of duty. In July 1918 the war was far from won and Mannock knew that he had work still to do in the training of his pilots; again, the very reason for the flight with Inglis that morning was to boost the young New Zealander's morale; Mannock would never have chosen such a time to literally commit suicide.

There have been only two published eyewitness accounts of Mannock's last flight: Donald Inglis and Edward Naulls, the latter a private soldier serving in the 2nd Essex Regiment which was occupying the British trenches in that part of the front line. Naulls wrote that after shooting down the two-seater, Mannock dived to 40 feet, just clearing the tallest of the remaining trees in Pacaut Wood. Inglis was at 100 feet. There was a great deal of rifle fire concentrated on the two SE5as and a machine gun commenced firing from 'near Robecq,' using tracers. Naulls stated that these hit Mannock's SE in the engine and a bluish-white flame appeared which spread rapidly, to the engine and cockpit areas. Mannock then made a *'right-hand turn and came toward our line.'* Just before the stricken SE reached the front lines, 'it turned *left, and back towards Pacaut Wood.'* Naulls then related that the SE continued to gradually loose height in a *'long glide,'* until it finally crashed in the general direction of Merville. (Author's italics throughout)

On close examination this statement does not make sense. The front-line trenches ran between the last of the trees in Pacaut Wood and the La Bassée Canal, which was at the backs of the British troops. The farthest part of the wood still in enemy hands was only three quarters of a mile from the British front-line trenches. From the northern-western edge of Pacaut Wood the front line then curved up towards St.Floris, passing to the east of Marquois and Robecq, leaving both villages behind the British front line. If Mannock had turned right, as stated by Naulls, then he would have been flying away from the nearest British trenches. If he had then turned left he would be flying away from Pacuat Wood, not back over it. The only way that Naulls' statement is valid, and for

Mannock's second turn, to the left, would put him back over the wood, is if by 'turn' Naulls meant that, in the first instance, Mannock actually turned through 270 degrees before turning left. But most people in using the term 'turn' mean a change of direction of 90 degrees – as Naulls would have had to have meant in describing Mannock's second turn. It could be argued that Naulls was referring to his left and right and not Mannock's, but this is at variance with Inglis' combat report, which clearly states that after being hit Mannock's SE turned to the right.

The question is: if Mannock were still in control immediately after his SE was hit, why should he have turned at all, when a straight glide of three quarters of a mile would have taken him safely over the British front-line trenches. The statements of Naulls and Inglis both agree that the flame first appeared on the right-hand side of Mannock's machine, but it is doubtful that this could have come from a machine gun 'near Robecq.' Robecq is two and a half miles from Pacaut Wood. It is more likely that Mannock's SE was hit by fire from the German front-line trenches or from positions in Pacaut Wood itself.

Inglis' combat report and later statement are both more feasible than Naulls' account. In his combat report Inglis stated: 'I then turned and followed Major Mannock back at about 200 feet. About half-way back I saw a flame come out of the right-hand side of his machine, after which he apparently went down out of control.' In a later statement, Inglis said that after the appearance of the flame the nose of Mannock's SE dropped slightly and 'he went into a slow right-hand turn, about twice and hit the ground in a burst of flame. Neither in his combat report, nor in his later statement was his height as mentioned by Naulls. From Inglis' combat report and later statement, it is clear that Mannock's SE was hit by ground fire, went at once into a slow right-hand spiral and crashed just south of La Pierre-au-Beure, only just over a mile from Pacaut Wood.

The theory advanced by James Dudgeon, one of Mannock's biographers, that Mannock shot himself to escape the pain of burning to death, does not bear examination. Dudgeon argues that the fire was on the left-hand side of Mannock's SE and that Mannock turned to the left to escape the flames and give him time to take his revolver from its pocket on the right-hand side of his cockpit. But the effect of a turn to the left would be that the slipstream would push the flames towards the body of the aeroplane. In addition, both Inglis and Naulls agree that the flame first appeared on the right-hand side of the SE. The fact that Mannock turned to the right would again have had the effect of pushing the flames towards the aeroplane. No pilot of Mannock's experience would have done this, and it argues that Mannock was not in control of the SE when it went into the slow right-handed spiral reported by Inglis. Again, the SE was at a height of less than 50 feet. If he were still conscious, Mannock would have had every chance to put the SE into a side-slipping turn to push the flames away from the cockpit area and make an attempt to land.

The known facts would all point to the conclusion that everything happened very quickly: from the first groundfire hitting the SE5a, possibly wounding Mannock so that he lost consciousness, to the burst of flames from the side of the SE, and the following uncontrolled right-hand spiral resulting in the fatal crash, only a short distance from Pacaut Wood.

It seems clear that Mannock did not burn to death in the air, or in the crash of his SE. His pocket notebook, identification discs, and revolver were anonymously returned to his family from Germany after the war. None of the effects showed any signs of having been in an intense fire and Mannock's notebook, which would have been in his tunic pocket, was only slightly browned around the edges of the pages. No details have ever been given of who sent Mannock's personal effects. It seems hard to believe that an unknown German soldier, in the chaos of post-war Germany, could have traced the address of Mannock's family and returned them, especially a revolver. It is logical to assume that some official agency was involved.

For 20 years after Mannock's death, Jim Eyles attempted to find the exact site of his friend's grave. The Imperial War Graves Commission could tell him only that German sources gave that Mannock had died on 26 July 1918. A letter from the Commission, dated 2 June 1921, to the Reverend E. Rogers of the Church Lads' Brigade, was more informative, stating that a report from the German Red Cross gave the grave's position as a point 300 meters north-west of La Pierre-au-Beure on the road to Pacaut, but regretting that the officers of the Commission had never been able to find the grave itself. This is almost certainly because they were looking in the wrong location, due to an incorrect translation of the German report or an error in the report itself. There is no road to Pacaut north-west of La Pierre-au-Beure and any burial of Mannock in that direction would have meant that the German troops would have had to carry the body several hundred yards towards the British trenches. However, there is a road to Pacaut north-east of La Pierre-au-Beure and the body of an unknown British flying officer, stripped of his personal effects and identification discs, had been found at that point, later buried in Laventie Military Cemetery, Grave 12, Row F, Plot 3, and marked 'Unknown British Aviator.'

After the war, many of Mannock's friends and comrades who admired his courage, leadership, and personal example, were determined that his many achievements should not go unrecognized; that if any airman deserved Britain's highest award for bravery, then that airman was Mick Mannock. They combined their forces in a long campaign for a posthumous award of a Victoria Cross. Their efforts finally came to the attention of Winston Churchill, the Minister of Air in the post-war government. Churchill instigated an inquiry into their claims and on 11 July 1919, the award of a Victoria Cross was gazetted to Major Edward Mannock MC and Bar; DSO and two Bars. After recording some of Mannock's aerial victories it concluded:

'This highly distinguished officer, during the whole of his career in the Royal Air Force, was an outstanding example of fearless courage, remarkable skill, devotion to duty and self-sacrifice, which has never been surpassed.'

Major James Thomas Byford McCudden

James Thomas Byford McCudden was born on 28 March 1895, the second son of William and Amelie – née Byford – McCudden. As was the case with Edward Mannock's father, William McCudden was serving in the regular army when James McCudden was born, and he also had served in the Egyptian campaign of 1881–1882. But there the resemblance ends. Unlike Edward Mannock, James McCudden enjoyed a happy and secure childhood as part of a large and loving family.

Both James and his elder brother William attended the garrison school at Brompton Barracks, Gillingham, Kent. James did well at the school, excelling at sport and, in particular, marksmanship – perhaps the first indication of the remarkable marksman he was later to become. He was also conscientious and extremely thorough in everything he did: an approach to life which would later typify his career as a pilot.

In 1909 the McCudden family moved to Sheerness and it was during the many hours spent watching the flying at nearby Eastchurch aerodrome that the two boys first felt the burning ambition to fly, however distant the realization seemed.

James McCudden left school when he was 14 years of age and took a temporary job with the Post Office as a messenger boy; temporary because he would not be eligible to enlist in the Royal Engineers until he was 15.[1] On 26 April 1910, a month after his 15th birthday, McCudden became No.20083, Boy McCudden, Royal Engineers. It was the humble beginning to a dazzling career.

In 1911 McCudden was serving on the Rock of Gibraltar and during his 18 months there he developed his skill with the rifle – tying with his captain for top score in one contest – and his ever present interest in aviation. James and William had never forgotten their days at Eastchurch, and when the Royal Flying Corps was formed in 1912, William lost no time in transferring from the Royal Engineers into the newly-formed Corps. In those early days there was some opposition within the RFC to the proposition that NCOs should be trained as pilots, but this was overcome; on 13 August 1912 William McCudden was the fourth NCO in the RFC to win his Royal Aero Club Certificate and he became one of the most respected pilots in the Corps. When James returned from Gibraltar in 1912 he was anxious to follow in his brother's footsteps and he at once applied for a transfer from his regiment. James McCudden entered the ranks of the RFC on 28 April 1913 as Air Mechanic Second Class No.892, reporting for duty to the RFC depot at Farnborough on 9 May 1913. He was a month past his 18th birthday.

The younger McCudden's career in the RFC nearly came to a premature end on his very first day at Farnborough. He was shown how to swing the propeller of a Caudron training aircraft and he conscientiously practiced this until the mid-day break. After lunch, he returned to the shed to recommence his practice. Unbeknown to McCudden, some unknown hand had been tampering with the controls of the Caudron: at the first swing the engine burst into life and the Caudron started to lurch out of the hangar, with McCudden having the presence of mind to throw himself flat just as the lower wing of the runaway passed over him. A Maurice Farman was blocking the path of the Caudron, but it was determined to escape the hangar and made short work of the Farman, its propeller destroying the tail booms and almost an entire wing before swinging round and attacking the Commanding Officer's car. McCudden picked himself up off the floor of the hangar and managed to switch off the engine of the Caudron, but not before it had severely damaged the car.

Sergeant Major Hudson fell in two men and marched the crestfallen McCudden to the guardroom, where he spent the next five days ruminating on his fate. He felt sure that he would be sent back to his regiment. When he was released under open arrest, McCudden went down to the aerodrome and begged Lieutenant B.T. James for a flight in his silver-doped BE2c, probably reflecting that it would very likely be the nearest he would ever get to being a pilot. Lt. James, 'with his usual good nature,' agreed. It was McCudden's first flight.

To answer for his crime, McCudden was finally brought before Lt. Col. F.H. Sykes, Commanding Officer of the Military Wing of the RFC. Sykes was a kindly man: no doubt influenced by both the obvious remorse of the young McCudden, and the respect with which the culprit's elder brother was held throughout the RFC, Sykes did not return McCudden to his regiment but gave him a severe reprimand, seven day's detention, and docked him 14 days pay.

When war came in 1914, both Sergeant William McCudden and Air Mechanic James McCudden were with 3 Squadron, which arrived at RFC Headquarters at Amiens in northern France on 14 August. During the next four weeks the Squadron never remained at an aerodrome for longer than four days. On 25 August the retreat of the British Army from Mons began, and 3 Squadron moved ten times in as many days, with Air Mechanic McCudden flying to the first of its new aerodromes with Lt. Joubert in an 80hp Blériot. During the retreat, the aeroplanes of 3 Squadron became scattered, often separated from their transport and mechanics. McCudden, and those other mechanics who flew with their pilots at each move, were often the only groundcrew available to service the machines and they worked hard and long, often into the hours of the early morning. Like their pilots, they slept where they could,

1. The writer has had the pleasure of reading the manuscript of McCudden's autobiography, now held in the Royal Air Force Museum at Hendon. Written in pencil, in an exercise book, in a clear, legible hand, the text shows hardly any alterations or spelling mistakes – mute testimony to the standard of education in the Army schools of the era.

James Thomas Byford McCudden.

but luckily the weather was fine and dry and McCudden often spent the night under the wing of his pilot's aeroplane or in the cockpit, covered with a waterproof sheet.

The retreat over, 3 Squadron was again based at St.Omer and in November James McCudden made his first war flight: a reconnaissance from St.Omer to Calais with Captain Conran, McCudden armed with a 0.303 rifle.

McCudden's keenness and expertise with both the engines and rigging of the machines of 3 Squadron did not pass unnoticed and he was promoted to Corporal in November 1914. On 1 April he was again promoted, to sergeant, and put in charge of all the engines in his Flight.

In early 1915, William McCudden had been posted to Home Establishment and in April was serving with 13 Squadron, which was forming up at Fort Grange, Gosport. On 1 May 1915, while giving an instruction flight in a Blériot trainer, William McCudden was killed. The cause of the accident was found to have been the engine choking on an excess of petrol, the loss of power causing the Blériot to stall and crash from 150 feet. This was a strange coincidence in view of James McCudden's own fatal crash in 1918.

In France, Jimmie McCudden was hit extremely hard by his brother's death. He was called to the orderly room and given the telegram by the recording officer, but he could hardly take in the news and read it again behind a hangar, where he was later found distraught with grief by one of the NCO pilots. Jimmie had always looked up to and admired William; the loss made him more determined than ever to emulate his brother and become a pilot and he put in an application for flying training. In June, McCudden and another sergeant, Webb, were recommended for a course of flying. Webb was sent for training, but McCudden was held back on the grounds that he was too valuable as a mechanic to be spared. His consolation was that he was selected to fly as a regular observer.

In July 1915, McCudden was given ten days leave. He recalled in his autobiography: '...which was very welcome, as I had been in France since August 14 1914 and we never had a holiday, not even on a Sunday, for the necessities of war prevented us from having too much spare time.'

On returning from leave, McCudden began to fly regularly as an observer. It was the early summer of 1915 and by August the Fokker E-type monoplanes, in the hands of the emerging German fighter pilots, began to

A Fokker E.III; 'A long brown form fairly streaking across the sky.'

take their toll of the RFC reconnaissance machines. On 15 December, 3 Squadron lost Lt. A.V. Hobbs and 2Lt. C.E.G. Tudor-Jones, both killed in action over Valenciennes by *Ltn.* Max Immelmann for his seventh victory, and the Squadron decided that for mutual protection three machines would fly the next reconnaissance in the area.

On 19 December three Moranes left the ground on a reconnaissance to Valenciennes, McCudden flying as a gunner/observer in Morane No.5088 with Lt. Saunders as his pilot. The three machines were five miles into enemy territory when the first Fokker was seen: 'a long brown form fairly streaking across the sky.' The Fokker pilot made several passes at the British machines, then turned his attention to Saunders and McCudden, coming towards them nose on and slightly above. McCudden could not fire at the Fokker while it was in that position but he coolly waited until it passed over the right wing of his Morane. Putting the Lewis gun to his shoulder, McCudden fired a long burst. The Fokker carried on and was lost to view. The Moranes were now over the German aerodrome at Douai and, looking down, McCudden saw that several enemy machines were leaving the ground. While McCudden was watching these machines he noticed that the Fokker had reappeared and was climbing up under the tail of the Morane. McCudden waited his moment, then told Saunders to turn and fired half a drum of Lewis at the Fokker from 300 yards range. 'The Fokker seemed rather surprised that we had seen him, and immediately turned off to my left rear as I was facing the tail.'

The Fokker pilot climbed to 300 feet above Saunders and McCudden before putting down his nose to fire. McCudden was waiting. As the Fokker dived, he fired again, forcing the enemy pilot to pull out of his dive and retreat to 500 yards from the British machine. Each time the Fokker pilot attempted to attack again, McCudden kept him at bay with a short burst of fire. The British machines flew on and carried out their reconnaissance of the Valenciennes area, still followed by the Fokker. On their return journey the Fokker finally left them, gliding down to its aerodrome at Douai. On his return McCudden wrote his combat report, the first of over a hundred.

That same afternoon McCudden flew another reconnaissance, this time with his commanding officer, Major Ludlow-Hewitt. The Major had just finished his own work and McCudden was sketching in the position of a new trench, when something told him to look up. A Fokker was about to attack the Morane. 'I happened to glance round and saw that beast Immelmann in his beastly Fokker about a hundred yards behind us, and a little higher, preparing to dive. I do not believe I ever moved so quickly in all my life. I dropped my maps on the floorboard and grabbed hold of my machine gun and fired at him from the shoulder, as the machine guns we use are very light and we use them in the machines just like rifles, and also the position he was in I could not use the gun in its proper mounting.'

Hearing McCudden firing, Ludlow-Hewitt put the Morane in a spiral dive. The Fokker followed but McCudden kept his nerve. 'In one position he passed over us and I was able to get some good shots in at him, and we then had the satisfaction of seeing a large piece of paper or map fall from the Fokker.'

No doubt shaken by such firm resistance, the Fokker pilot flew away. McCudden concluded that if the pilot of this Fokker, and the one which had attacked him in the morning, was the dreaded Immelmann, 'Then he would not fight even when the odds were even,' and completely lost his fear of the German ace.

In January 1916 James McCudden was awarded the French *Croix de Guerre,* his first decoration, but even more satisfying was the news that he was to train as a pilot. On 1 February McCudden was promoted to flight sergeant and left for England for flying instruction.

At Farnborough on 22 February 1916, a cold, windy morning, James McCudden began his pilot training with a 20-minute flight in a Henry Farman pusher. By this date McCudden had over a hundred hours of flying time as an observer and was thoroughly at home in the air, and his instructor, Sergeant-Major Power, soon realized that he had a very apt and capable pupil. On 9 April, McCudden was posted to Fort Grange, Gosport – the scene of his brother's fatal crash – and after four hours of dual instruction he finally flew solo a week later. Later that day, McCudden took the test to qualify for his Royal Aero Club Certificate, No.2745.

McCudden's hopes that he would now be sent to France as a scout pilot were dashed when he was posted to the Central Flying School as an assistant instructor. RFC policy at this time was for the better pupils at the elementary flying schools to be 'student instructors' and McCudden took up his first pupil when he himself had only eight hours flying time.

After staying at the CFS for just over two months, McCudden was posted to France. But his ambitions to be a scout pilot were again frustrated. On 8 July he was posted to 20 Squadron, which was equipped with the FE2b, a large, ungainly pusher aeroplane used for general reconnaissance and artillery spotting work. McCudden, however, had now been a professional soldier for seven years and he accepted his orders with good grace. In the event he did not have long to wait. On 3 August, after landing from giving his two mechanics a joyride around the aerodrome during a test flight, he was told that he was to leave at once for 29 Squadron. He was 'very glad that I was to fly scouts in France.' 29 Squadron was equipped with the DH2, a light, compact little pusher which had also equipped 24 Squadron, the RFC's first fighter squadron.

McCudden flew with 29 Squadron for seven months, gaining a great deal of invaluable combat experience. By the time his service with 29 Squadron came to an end he had scored five victories – three enemy machines destroyed and two out of control – and had damaged several others. It is possible that, while serving with 29 Squadron, McCudden gained a distinction of which he was blissfully unaware. Recent research has shown that he was almost certainly claimed by Manfred von Richthofen as his 15th victory.[2]

On 27 December 1916, von Richthofen claimed a Vickers two-seater shot down at Ficheux, south of Arras, at 4:15p.m. (German time). In his combat reports throughout 1916 and 1917, von Richthofen indiscriminately referred to any pusher aeroplane with twin tail booms as 'Vickers,' but these were the two-seater FE2 or the single-seater DH2 and FE8, which had replaced the Vickers Gunbus flown by the RFC. There were no losses of FE2s on 26 December which fit for the time and place of Richthofen's claim, but three DH2s of 29 Squadron were in combat with Albatros scouts at the time and location in question.

A Flight of six DH2s had left the ground at 2:00p.m., but by the time they had crossed the lines near Arras at 10,000 feet, three of their number had dropped out with various technical troubles. The remaining three DH2s – Captain H.J. Payn, Lt. A. Jennings, and James McCudden – attacked five Albatros scouts at approximately 3:15p.m.

McCudden's Bristol Scout serial 5575. McCudden flew this Scout while stationed at Dover in early 1917. *Teddie,* carried in front of the cockpit, was the name of a current girlfriend, Teddie O'Neil, a dancer in a West End show.

(British Time). McCudden attacked an Albatros which was firing at Jennings; the enemy machine turned at once and came at McCudden head on. McCudden returned the fire as the two aircraft closed, but his Lewis gun stopped after 20 shots with a double feed. Unable to rectify this while in the middle of the enemy scouts, McCudden half-rolled away and dived, keeping the DH2 vertical for a few hundred feet. When he came out of his dive, the Albatros was now on his tail, still firing. McCudden half-rolled away again, but the German pilot was persistent, following him down. McCudden continued his evasive manoeuvres until he was at 2,000 feet, but the Albatros was still there, finally forcing McCudden down to 800 feet. At this height the Albatros came under fire from the British anti-aircraft fire and climbed back to join its companions.

When McCudden returned to base he was told that Captain Payn had last seen him going down out of control 'over Hunland with a fat Hun in attendance,' and it was assumed that he had been shot down.

This combat of the 29 Squadron DH2s fits for time and place with von Richthofen's combat report, with the exception that Richthofen states that the 'Vickers' he had fought – which he described as 'very courageously flown' – had crashed and that it was a two-seater. In the heat of the combat Richthofen may well have thought that he was fighting a 'Vickers' two-seater, a type he was used to meeting, although it is strange, given the duration of his fight with McCudden – if McCudden it was – that he did not realize his opponent was flying a single-seater. Richthofen's claim that the 'Vickers' had crashed could have been wish fulfillment on his part: he had fought the enemy machine down from 10,000 feet to within 800 feet of the ground, with it falling all the way. It is perhaps excusable that he considered it must have crashed.

McCudden returned to England on 23 February 1917 as a Second Lieutenant, and on 12 March he was awarded a Military Cross for 'conspicuous gallantry in action.'

2. *Under the Guns of the Red Baron.* Franks, Giblin, and McCrery. Grub St. London. 1995.

McCudden's next posting was as a fighting instructor to the 6th Training Wing at Joyce Green in Kent. While lecturing and demonstrating combat techniques at the school he met Edward 'Mick' Mannock, and the two pilots became great friends, a friendship cemented by their socialist ideals.

McCudden was next posted to Dover and while there was allocated his own personal aeroplane, a Bristol Scout. When McCudden's time at Dover came to an end he was given the task of lecturing to all the training squadrons of 7th Wing in East Anglia, and he was given the use of a Sopwith Pup, A.7311.

The German Gothas raided the English mainland for the first time on Friday 25 May 1917, following this with two minor attacks on Sheerness and Shoeburyness on 5 June before mounting a large raid on London on 13 June. On 3 June, after the raid of 25 May, McCudden had had A.7311 armed with a Lewis gun to fire over the top wing in anticipation of being able to get amongst the Gothas when they next attacked, and he started to fly standing patrols in the hope of intercepting them.

On 13 June McCudden caught 16 Gothas on the way back to their base, well out to sea over Shoeburyness. McCudden closed the range, came up underneath the nearest bomber and fired a complete drum of Lewis into it, but the Gotha simply swerved a little then carried on its course. The bombers steadily drew away from the Pup and McCudden was finally forced to let them go, over 20 miles out from the coast.

On 7 July, McCudden engaged another formation of Gothas, this time over Tilbury. He dived on the last machine in the German formation and opened fire at 100 yards, holding down his trigger until within 50 yards, firing a whole drum of Lewis. McCudden pulled the Pup out of its dive so violently that the seat bearers broke. Undeterred by this, McCudden changed drums and attacked twice more, but each attack was unsuccessful. In view of McCudden's excellent marksmanship it was remarkable that these attacks gained no decisive result, but he ran out of ammunition and he contented himself with shadowing the enemy formation for 25 minutes, hoping to distract their attention from any other British aeroplanes which might attack them. In a careless moment McCudden allowed the Pup to slip out of the blind spot of the nearest Gotha and an accurate burst of fire from its gunner put a bullet through the windscreen of the Pup. McCudden decided that the gunner was 'a very nasty man' and spun the Pup out of the action, causing the Gothas to claim they had shot down one of the defending aircraft.

On 12 July McCudden went back to France on a refresher course, joining 66 Squadron, based at Estrée Blanche and equipped with Sopwith Pups. Over the next three weeks McCudden flew 22 operations with 66 Squadron, his logbook indicating that he claimed one victory, which was classified as indecisive and not allowed. On 21 July, McCudden was invited by Major Richard Blomfield, commanding 56 Squadron, also based at Estrée Blanche, to fly an evening patrol in one of the Squadron's SE5as. In company with Captain Prothero and Lts. Gerald Maxwell and Arthur Rhys Davids, McCudden took off at 6:30p.m. Prothero dived to attack a formation of Albatros scouts over Houthulst Forest and McCudden, unused to the speedier SE5a, was left behind. When he finally rejoined the Flight he shot at a silver-gray Albatros, which 'turned east and wobbled laterally' and McCudden watched the enemy machine as it fell in a slow spiral, with one wing right down, from 15,000 to 6,000 feet. He then came under attack from other enemy machines and was forced to clear. When McCudden landed back at Estrée Blanche he did not claim the silver-gray Albatros – in his opinion it was still under control when he last saw it – but he knew that somehow or other he must arrange a posting to 56 Squadron to fly the SE5a and he made no secret of his wish to Major Blomfield.

Blomfield was only too eager to have McCudden posted to his Squadron as a Flight Commander. Not only was he impressed by McCudden's keenness and dedication, but one of his own Flight Commanders, Captain G.H. 'Beery' Bowman, who had flown with McCudden in 29 Squadron, highly recommended him. Only eight days after returning to England, McCudden was told he was to return to France as a Flight Commander in 56 Squadron. He wrote: 'I don't think I have often experienced such pleasure as when I was able to call myself a Flight Commander in No.56 Squadron.'

McCudden arrived at 56 Squadron in the evening of 14 August 1917. The next morning he spent several hours adjusting and firing his guns at the target butts before he was satisfied that they were correctly aligned. When he went into the Mess for lunch there was a certain amount of ironical booing from the other pilots, who considered his morning's work as being that of a show off. They had yet to realize McCudden's absolute thoroughness in ensuring that his guns and aeroplane were exactly right. As a professional soldier he knew that this was the only way to fight, but to the other pilots he was just another Flight Commander – no more, no less – who had scored his five victories – fewer than some of his Flight – in the now laughably obsolete DH2. Gerald Maxwell, one of the pilots present that morning, was later to ruefully remark, 'We just didn't realize Mac's worth at the time.'

McCudden scored his first victory with his new squadron on 18 August, shooting down an Albatros, possibly from *Jasta* 14, east of Houthem. The next day he shot down another Albatros, an all red machine with yellow stripes, over Gheluvelt at 5:25p.m. He continued with this run of successes on 20 August with yet another Albatros scout, which fell in flames to the south-east of Polygon Wood at 6:50p.m., following this ten minutes later with another out of control. These four victories in three days earned McCudden the respect of his fellow pilots, but it was to be nearly a month before he scored again. Despite his care and attention to detail, he had continual trouble with his guns, losing many probable victories by their jamming at the crucial moment. The extent of his frustration is evident in a letter to his father on 10 September.

'I have had a lot of trouble with my guns jamming of late, and have lost a lot of Huns over it...This morning I got 20 yards behind a Hun and took a sight on the back

McCudden's famous SE5a B4891. This SE5a was fitted with a spinner, painted red, taken from an LVG which McCudden had shot down on 30 November 1917.

of his neck and pulled the triggers, and neither gun fired a single shot, and while I was rectifying the stoppage the Hun got away.'

Four days later McCudden's run of bad luck ended. In a fight with Albatroses from *Jasta* 10 he attacked *Oberleutnant* Weigand, slightly wounding him and hitting his Albatros in the engine. The Albatros began to vibrate violently and Weigand shut off his engine and went down in a flat spiral.

A period of bad weather followed this victory, but on 19 September McCudden, flying alone, found a DFW two-seater over Merville. On McCudden's attack the DFW dived away, so violently that a small black box and a number of signal cartridges fell out of its cockpit. McCudden followed it down to 10,000 feet before leaving it and climbing away to regain his height. Thirty-five minutes later McCudden saw a Rumpler crossing the lines over Armentieres at 14,000 feet. He carefully stalked this Rumpler, getting into the sun and waiting until the enemy machine was well west of the lines before making his attack. After one shot, McCudden's Lewis gun stopped with a double feed and he switched to the slower firing Vickers gun. The Vickers fired 30 shots then also stopped. McCudden's feelings can be imagined: after good tactics and making his attack from such a favourable position, he was again to be robbed of a result by faulty guns. But his aim had been good. The propeller of the Rumpler jerked to a stop and it began to go down, leaving a trail of blue smoke. McCudden cleared his Vickers of the stoppage and attacked again, firing ten shots. At 4,000 feet he could see that the enemy observer had slumped down into his cockpit, leaving his gun unattended. McCudden was determined to crash this Rumpler on the British side of the lines and he attacked again, but the strong westerly wind had blown the fight to the east and the Rumpler finally crashed a mile behind the German front-line trenches. This victory was witnessed by pilots of 10 Squadron.

On 21 September, McCudden shot down another two-seater, over Gheluwe, bringing his victory total to 12. Leading 'B' Flight on an evening patrol on 23 September, he shot down a DFW – crewed by *Ltn.* Gustav Rudolph and *Uffz.* Rudolph Franke of *Fl.Abt.* 6 – over Houthem at 6:00p.m., before leading the Flight into the classic fight with *Ltn.* Werner Voss, *Staffelführer* of *Jasta* 10. After a fight lasting ten minutes, Voss was finally shot down by Lt. Arthur Rhys Davids of 'B' Flight.

McCudden took off alone on 27 September to look for two-seaters. He saw an LVG shoot down a French SPAD, then closed with the enemy machine over Houthulst Forest and shot it down in flames. The enemy observer, *Uffz.* Hans Gossler, jumped out of his burning machine and fell in the German lines, but the LVG and its pilot, *Uffz.* Bruno Weiderman of *Schutzstaffel* 27, both crashed in the British lines, the first of McCudden's victories to fall on the British side. At 8:00a.m. the following day McCudden led his Flight into action with a formation of five Albatros scouts over Houthulst Forest and shot one of them to pieces, the pilot – possibly *Ltn.* Pastor of *Jasta* 29 – falling out of his aeroplane. Over breakfast that morning, Lt. Richard Maybery wryly commented that McCudden was becoming expert at turning Germans out of their aeroplanes.

McCudden opened his scoring in October on the first day of the month, shooting down a Rumpler over Herlies at 2:50p.m. On the evening of 6 October 56 Squadron held a dinner to celebrate the awards of a Bar to McCudden's MC, a second Bar to Barlow's MC, and a DSO to Lt. Rhys Davids – all 'B' Flight members.

After a period of bad weather, during which there was

McCudden is second from right, front row, in this photograph of instructors and staff of the No.1 School of Aerial Fighting at Ayr in 1918. Gerald Maxwell (27 victories), who flew with McCudden in 56 Squadron, is the first officer on the left, front row. Fifth from the left, front row, is the CO of the school, Lt.Col. L.W.B. Rees, VC. Captain A. Dover Atkinson (eight victories) is seated third from the left in the second row.

no flying, McCudden commenced scoring again on 17 October with a victory over an LVG. McCudden's shots set fire to the center section of the LVG, the fire burning the fabric off the rudder. The fire then went out, possibly due to the petrol in the center section being exhausted, and the enemy pilot put his damaged machine into a glide towards the British trenches. McCudden later wrote: 'he had no means of turning either way and I was interested in following him down until he landed in our lines, for we were now over Vlammertinghe (sic), which was 15 miles from the trenches.' Lts. Barlow and Muspratt now arrived and, not noticing McCudden, attacked the LVG, which went into a steep dive, both wings breaking off before it crashed in a field near Vlamertinghe. McCudden followed the LVG down and landed in the same field. Leaving his engine ticking over McCudden went to the first of two groups of infantry which had gathered around the wreckage of the enemy machine. The first group was standing over the body of the enemy observer, 'a huge man named Ernst Hadrich...' Seeing that Hadrich was dead, McCudden went over to the second group, milling around the remains of the LVG, a 100 yards away. The Australian infantry had already stripped the wreckage of souvenirs and McCudden was horrified by the unedifying way in which the troops were behaving. This LVG C.V was from *Fl.Abt.* 8 and was crewed by *Oblt.* Ernst Hädrich and *Ltn.d.R.* Heinz Schuberth, both killed.

On 21 October McCudden shot down a Rumpler from *Fl.Abt.* 5, which fell in the British lines at Mazingarbe. This victory came after a long and careful stalk of the Rumpler and was the first example of the art which McCudden was to perfect in the coming months – the stalking and destruction of high-flying German reconnaissance aeroplanes. Two days after this victory – his 18th – McCudden went on leave.

McCudden returned to France on 10 November. During his leave the Squadron had moved to Laviéville, was now under the orders of 13th Wing, and had a new commanding officer, Major Balcombe Brown. On 18 November McCudden recommenced his scoring during his first patrol over the new front, shooting down a DFW over Bellicourt. Five days later, on 23 November, he shot down *Vzfw.* Karl Bey of *Jasta* 5. Bey's Albatros 'went down and hit the ground, in a vertical dive with the engine full on, a fearful whack,' near Rumilly.

Bad weather stopped flying until 29 November, but McCudden made up for the period of enforced inactivity by shooting down two two-seaters: the first, a DFW C.V of *Fl.Abt.(A)* 202, flown by *Ltn.d.R.* Kurt Dittrich and *Ltn.d.R.* Manfred Hoettger, in the morning at 7:30a.m., south of Bellicourt; the second at 1:00p.m. over Rouvroy: another DFW C.V, crewed by *Ltn.d.R.* Georg Dietrich and *Ltn.* Dietrich Schenk of *Fl.Abt.* 268. The first DFW broke up in mid-air and 'literally fell to a thousand pieces.' and the second DFW also broke up, all four wings breaking off.

The following day, McCudden shot down yet another LVG, this time from *Fl.Abt.* 19. The LVG C.V No.9458/17, crewed by *Vzfw.* Flohrig, who was fatally wounded, and his observer, *Gefr.* Eckerle, who was unhurt and taken prisoner, crashed in the British lines and was given the number G.94.

December was to be an extremely successful month for McCudden. He began on 5 December by shooting down a Rumpler from *Fl.Abt.* 54, crewed by *Ltn.* Fritz Pauly and *Ltn.* Ernst Sauter, following this the next day with two more victories: a photographic Rumpler from *Fl.Abt.* 255 in the morning and at 3:00p.m. a blue-tailed Albatros D.V from *Jasta* 5, out of control over Fontaine. McCudden also attacked a Hannover CL – the first of the type he had seen – and drove it down damaged, but he made no claim for a victory over the new type.

McCudden met his match on 12 December. After driving away some enemy scouts which were firing into the British trenches, 'B' Flight spotted a DFW just below the clouds over Hermies, four miles west of the British lines. McCudden attacked the DFW alone and fought it

McCudden at Ayr with his personal Bristol M1c C.4955.

for five minutes, from 4,000 to 500 feet, but the crew were very experienced and McCudden broke off the combat, discretion being the better part of valour. '...the Hun was too good for me and shot me about a lot. Had I persisted he certainly would have got me, for there was not a trick he did not know...' The disappointment at not downing this DFW was perhaps softened by a telegram during the day from General Trenchard, congratulating McCudden on the award of a DSO.

The crew of a Rumpler attacked at 11:05 on 15 December were not so skillful. McCudden attacked this Rumpler over Villers and shot it down to crash just to the east of the Bois de Vaucelles. McCudden failed to score again until 22 December. Taking off alone, he flew to the area of St.Quentin to look for enemy two-seaters. Visibility was good and McCudden knew he would not have long to wait before an enemy photo-reconnaissance machine came over the lines. McCudden waited at 17,000 feet. His patience was soon rewarded: a pair of DFWs flew directly underneath him over Holnon Wood. McCudden dived to attack the nearest DFW, 'disabled' it, and left it gliding down into the British lines before turning away to attack its companion. This DFW 'fought as though for his life,' but McCudden kept in his attacking position, his fire damaging it as it made for the safety of its own lines. Looking round, McCudden then saw that the first DFW had turned east and was also making for the safety of the German lines. McCudden left the second DFW and again attacked the first. He could not see an observer – the DFW was possibly flying without one – and McCudden tried to force the DFW pilot to turn to the west. The enemy pilot, *Uffz.* Anton Bode of *Schutzstaffel* 5, refused to turn so McCudden fired 'a lot more ammunition at him.' The DFW went down in a flat spiral glide and crashed in the British lines just to the south-west of St.Quentin.

The next day, 23 December, was McCudden's most successful day in a highly successful month. At 11:25a.m. he shot down an LVG over Anguilcourt, which fell in the German lines; 55 minutes later he shot the starboard wings off a Rumpler from *Fl.Abt.* 23 to send it crashing at Contescourt. After lunch, leading his Flight, McCudden shot down another Rumpler, this one from *Boghol* 7, which fell in the British lines at 2:40p.m., the crew being taken prisoner. McCudden's fourth victory for the day was a yellow-fuselaged LVG C.V from *Schutzstaffel* 12, shot down to crash near Metz-en-Couture. This was the first time a British pilot had destroyed four enemy aeroplanes in a single day and McCudden received many telegrams of congratulations.

McCudden was now seemingly invincible in the air. On 28 December he scored another three victories: a Rumpler from *Fl.Abt.* 7 at 12:15p.m.; another Rumpler, this one from *Fl.Abt.(A)* 40, fell in flames 15 minutes later over Flers; and at 12:55p.m. an LVG C.V from *Fl.Abt.(A)* 210 fell under his guns over Havrincourt Wood. Another combat with an LVG nearly resulted in a fourth victory for the morning, McCudden's fire causing a flare of flame to erupt from the fuselage of the LVG, and steam and water to pour from its radiator, but he was then forced to break off the combat as he was running short of petrol.

Not content with these successes, McCudden was in the air again the next day, shooting down an LVG from *Schutzstaffel* 10 into the British lines at Havrincourt at 10:00a.m., and after lunch, at 2:00p.m. north-east of Epéhy an LVG from *Fl.Abt.* 33. This LVG was McCudden's last victory in December 1917. It had been a remarkable month. The total number of victories scored by 56 Squadron during December had been 17; no less than 14 of these had been scored by McCudden. During December 1917, McCudden literally *was* 56 Squadron. At the close of 1917, McCudden's victories numbered 37, all but three confirmed by independent witnesses. There can be little doubt that at the end of 1917 McCudden was one of the most efficient destroyers of enemy aeroplanes that had yet been seen in the skies above France.

McCudden had several indecisive combats in January 1918, but he failed to score until the morning of 9 January, when he shot down an LVG over Graincourt. This LVG was 56 Squadron's 250th victory; it was also 'B' Flight's 100th, making it the first Flight in the RFC to achieve a century.

McCudden took off alone on the morning of 13 January and within 25 minutes had shot down an LVG, a DFW, and another LVG. McCudden's expertise and the seeming ease with which he was shooting down enemy two-seaters was the talk of the RFC and he was sent on a short tour around the squadrons to lecture on his methods. On his return to war flying, McCudden shot down an LVG from *Fl.Abt.(A)* 202 on 20 January and he followed this four days later with a DFW from *Fl.Abt.(A)* 240 over Monchy-le-Preux, fatally wounding Ltn. Pollocks. McCudden was the only member of 56 Squadron to have scored in over a month and the Squadron victory lists show his name for 16 consecutive entries. On 25 January, McCudden scored again, shooting down a Rumpler from *Fl.Abt.(A)* 225 over Urvillers at 2:40p.m.

The weather was bad for the next few days and McCudden took advantage of the enforced inactivity to carry out modifications of his own design to the engine of his SE5a. He tested the SE5a on the morning of 28 January and was delighted with the increased performance. On 30 January, flying alone, McCudden attacked a formation of four enemy scouts and shot down *Vzfw.* Adam Barth from *Jasta* 10, the Albatros D.V No.4565/17 going down out of control over Anneux at 11:15a.m. Turning away from Barth, McCudden attacked a Pfalz D.III and shot it down out of control. These victories were McCudden's first over enemy scouts since 6 December and were his last victories in January 1918.

February 1918 was to be McCudden's last full month in 56 Squadron, but it was to be second only to December 1917 in the number of enemy aeroplanes shot down, no less than 11 falling under his guns. He started the month's scoring on 2 February, with an LVG C.V from *Boghol* 7, shot down east of Vélu, but he failed to score again until 16 February, bad weather practically grounding 56 Squadron for 13 days. The weather having improved, McCudden took off on 16 February and his return to operations was an unlucky day for four German two-seaters. McCudden's first victory was a Rumpler from *Fl.Abt.(A)* 269, all its wings breaking off under McCudden's fire. Ten minutes later McCudden caught an DFW from *Fl.Abt.(A)* 202 which fell in flames before breaking into pieces over Le Catelet. Twenty minutes later, after sending a LVG down with water pouring from its radiator, McCudden caught another Rumpler, this one over Hargicourt, and shot it down. This Rumpler was confirmed by another pilot, but Major Balcombe Brown refused to allow it as a victory. Extremely annoyed, McCudden took off after lunch and shot down a Rumpler crewed by *Vzfw.* Zeuch and *Gefr.* Lechlieter from either *Schutzstaffel* 27 or 29 which fell at Lagnicourt.

The disallowed claim of the Rumpler of the morning was later confirmed by an anti-aircraft battery for McCudden's 50th victory. Balcombe Brown's initial refusal of McCudden's claim of this Rumpler, even though McCudden was a pilot of proven worth, is contrary to the recently expressed opinion in research journals that squadron commanders of RFC squadrons were lax in the confirmation of given victories.

McCudden shot down an LVG over Guémappe the next day, but his victories the following morning over two Albatros D.Vs from *Jasta* 35b gave him perhaps the most satisfaction. McCudden and 'B' Flight attacked a formation of Albatroses over Vitry-en-Artois and McCudden shot down two of their number. As the first burst into flames and went down inverted, McCudden saw that it had a green tailplane and carried a large letter 'K' on its top wing. Since arriving at Laviéville in November, patrols from 56 Squadron had often fought with formations of Albatroses, well led by a very skillful Albatros pilot, whose machine sported a green-painted tailplane and the letter 'K' on the top wing. The Squadron mistakenly believed that Richard Maybery had been shot down in December by this German pilot – who they had dubbed 'Greentail' – and McCudden was delighted that he had at last shot down their wily opponent and avenged Maybery, but he wasted no time in jubilation and attacked a blue-tailed Albatros from the enemy formation, shooting it down to crash at Quiery-le-Motte.

McCudden's first victory was not 'Greentail,' but *Uffz.* Julius Kaiser, who jumped from his burning aeroplane and was killed. The blue-tailed Albatros, McCudden's second victim, was flown by *Uffz.* von Stein. Stein was wounded but managed to crash-land his damaged aeroplane.[3]

McCudden's next victory was a DFW on 21 February, which he shot down in flames after firing only six rounds – it was with some justification that 'Beery' Bowman, the 56 Squadron 'C' Flight Commander, once described McCudden as 'a shooting genius.'

McCudden then went down with a very severe cold and while in bed he was visited by Lt.Col. Playfair, who informed him that he was to be posted to Home Establishment in the very near future, adding that he was to do no more flying in France before leaving for England. McCudden was annoyed at this decision: although he had been in France for seven months he still felt fresh and was full of confidence. To McCudden's frustration, the next morning dawned beautifully clear, cold and fine, and he telephoned Playfair and begged for permission to fly, strengthening his case by lying to Playfair that he felt much better. Playfair at last granted permission and McCudden took off alone at 9:45a.m.

Despite his assurance to Playfair that he was better, McCudden felt extremely ill: his throat was very sore and the cold and height were affecting it. He gained his height of 15,000 feet and attacked an LVG over Masnières. The LVG also had a large number '6' painted on its top wing, as did McCudden's SE5a, and the pilot was extremely accomplished, throwing the two-seater about like a scout. After five minutes of 'some very

3. 'Greentail' and his fellow pilots were from *Jasta* 5.

The aerodrome at Auxi-le-Chateau, the scene of McCudden's fatal crash on 9 July 1918. McCudden's SE5a crashed into the wood at the rear of the right-hand hangar of the three in line.

exciting and skillful fighting' the LVG, damaged by McCudden's fire, dived steeply east and escaped. McCudden now had time to look around for the first time since he had attacked the LVG and he noticed a formation of three Albatroses fighting with three SE5as from another squadron. McCudden climbed up into the fighting, but by the time he arrived the SEs had driven the Albatroses east.

McCudden then flew north and fought with two Hannover CLs, but they cooperated well and he could not force a decision. As the fire from the enemy two-seaters was accurate and hitting the SE, McCudden broke off the action and flew further north. He next saw a Rumpler returning to its own lines from over Arras and followed it to Douai before he was in range. After a burst from both guns the Rumpler burst into flames and fell in pieces to crash west of Oppy. McCudden then returned to the south and saw two enemy machines over Chérisy. Getting to within range he saw they were a DFW escorted by a Hannover CL. Seeing the SE, the DFW pilot dived for the safety of his own lines. The Hannover pilot turned first towards McCudden, but then turned away and followed the DFW. McCudden had attacked many Hannovers and although he had sent down several as damaged he had never succeeded in destroying one, and he now made, 'an instant resolve...I am going to shoot down that Hannover or be shot down in the attempt.' McCudden got into position and fired both guns until the Hannover fell to pieces, the wreckage fluttering slowly down as his 57th victory.

Turning north again, McCudden saw a Rumpler crossing the lines over Oppy and above the SE. McCudden climbed as hard as he could and slowly caught up with the enemy machine, but just as he was about to open fire another Rumpler flew across the nose of the SE. McCudden left the first Rumpler and attacked the second, opening fire at long range. Pieces came off the Rumpler but it continued to glide down over its own lines under control.

McCudden had now exhausted his ammunition, had been up for over two hours, and in combat nearly the whole time. 'That was one of the very busiest patrols that I had ever done,' he later wrote, 'But oh! as I flew back to the aerodrome I felt so ill that I thought I was dying. I just managed to stagger back as far as the aerodrome and land safely, and as I got out of my machine I was ready to drop. However, after I had rested and got warm, I was all right again.'

On reflection, McCudden came to the conclusion that the LVG with the number '6' on its top wing, which had given him such a good fight, had been a decoy, to enable the three Albatroses to surprise him and shoot him down. On 27 February he flew over to 84 Squadron to thank the pilots of the SEs who had engaged the three Albatroses. While there he cautioned his younger brother, Anthony, against his recklessness, advising him to be more careful.[4]

On 2 March, his posting to Home Establishment

4. At this time Lt. J.A. McCudden had eight victories. He was killed in action on 18 March 1918 in a combat with *Ltn.* Hans Wolff of *Jasta* 11.

McCudden's grave in the cemetary at Wavans in 1918.

imminent, McCudden was given a farewell dinner in Amiens. Representatives of all the squadrons in 3rd Brigade attended and the Commandant of Amiens. Over 50 officers sat down and many more crowded in after the dinner to give their best wishes and congratulations. On 4 March, 56 Squadron gave their own dinner in McCudden's honour and he was presented with a silver model of his SE5a.

McCudden left France and 56 Squadron on 5 March. His value to the Squadron can hardly be overestimated, his personal contribution to its success, unequaled. McCudden had taken command of 'B' Flight on 15 August 1917, and had scored 52 of the Flight's 77 victories for the loss of only four pilots. The safety of his pilots was of utmost importance to McCudden, this low casualty rate was an indication of the excellence of his leadership, and he later confided to his sister Mary that it gave him more pleasure than the number of his victories.

On 2 April 1918, James Thomas Byford McCudden was awarded the Victoria Cross. The citation read, in part, 'For conspicuous bravery, exceptional perseverance, keenness, and very high devotion to duty...' and ended, '...As a patrol leader he has at all times shown the utmost gallantry and skill, not only in the manner in which he has attacked and destroyed the enemy, but in the way he has during several aerial fights protected the newer members of his flight, thus keeping down their casualties to a minimum. This officer is considered, by the record he has made, and by the very great service he has rendered to his country, deserving of the very highest honour.'

After leave, McCudden was posted to the No.1 School of Aerial Fighting at Ayr, Scotland, taking up his new post on 10 April. In June he made a brief two-day visit to 56 Squadron and flew two patrols. Although he attacked several enemy aircraft during these patrols he had no success. Knowing from camera gun evidence at Ayr that his shooting was as accurate as ever, McCudden had the guns of the borrowed SE5a checked. The guns were badly out of alignment and a pilot in 56 Squadron at the time recalled that McCudden was 'hopping mad' and told off the unfortunate pilot of the SE in no uncertain terms.

Like many fighter pilots, McCudden was never happy for long away from active service and the comradeship of squadron life, and he pestered the authorities to send him back to France. In late June, McCudden, now promoted to Major, learnt that he was to take command of 60 Squadron. On the morning of 9 July, he visited his sister Mary at her office in London to say good-bye before returning to France. As he got into the taxi, which was to take him to Hounslow to collect an SE5a to fly to France and take up his new command, he turned to Mary and gave her a bulky envelope, saying 'I'd like you to look after these for me, sis.' Once back in her office, Mary opened the envelope. It contained all his decorations.

McCudden's grave today.

The afternoon of 9 July 1918 was hazy and visibility was limited. At 5:30p.m. an SE5a landed at Auxi-le-Chateau aerodrome, only five miles from 60 Squadron's aerodrome at Boffles. As the SE5a rolled to a stop the pilot beckoned over two corporals and asked them where he was and the direction of Boffles. They told him and helped turn the SE. With a quick glance at the windsock, the pilot taxied to the far end of the aerodrome and took off, in a steep climbing turn. The SE hit the trees on the edge of the aerodrome and crashed into the wood. One of the first to arrive at the crash was a rigger who had served with McCudden and he recognized the pilot, lying unconscious close by one wing of the wrecked SE. McCudden was rushed to No.21 Casualty Clearing Station but he died at 8:00p.m.

James Thomas Byford McCudden, VC, DSO and Bar, MC and Bar, MM, and *Croix de Guerre,* was buried on the afternoon of 10 July in the small military cemetery at Wavans, close by Auxi-le-Chateau. Pilots from 56, 60, and other squadrons attended the ceremony and General Salmond was present. The ceremony angered many of the pilots at the graveside. One commented, 'It was poorly arranged and rushed through.' Another was even more forthright in his condemnation. 'The ceremony made my blood boil – all in Latin, mumbled, so that even if one knew the language, he couldn't have heard it. Nothing human in it at all, and far from impressive. Richthofen – an enemy – had a far better funeral, and if anyone deserved a real memorial, it was McCudden.'

Chief Air Marshal Sir John Salmond later wrote a tribute to McCudden, ending with '...the secret of his remarkable success lay in the fact that he fought with his head as well as with his great heart.'

Lieutenant Alan Arnett McLeod

In November 1917, 2 Squadron, Royal Flying Corps, was based at Hesdigneul, equipped with the Armstrong Whitworth FK8, a sturdy but relatively slow two-seater, popular with those who flew it. The Squadron's main duties were reconnaissance and artillery spotting, but the Big Ack – as it was affectionately known – was also used for contact patrols, ground strafing, and bombing, the latter by both day and night.

On 29 November 1917 a new pilot reported to the squadron, an 18 year old Canadian: Alan Arnett McLeod. The young man from Winnipeg soon showed his aggressive spirit by strafing enemy trenches on the completion of each of his sorties, an activity which led his Commanding Officer, Major W.R. Snow MC, to describe him as 'a young fire eater.' It was vital work; the Battle of Cambrai had opened on 20 November 1917 and instructions from RFC Headquarters to the RE8 and FK8 squadrons were to aid the fighter squadrons in bombing and machine gunning enemy troops and positions at every opportunity.

On 18 December 1917, while flying an artillery shoot near Hulloch, McLeod saw an Albatros V-Strutter strafing the British trenches. Diving to the aid of the British troops, McLeod and his observer drove off the German fighter with fire from their guns. The following day, again flying a shoot by Hulloch, McLeod was attacked by eight Albatros fighters, coloured yellow with green stripes. The German Jasta was obviously out to revenge its humiliation of the previous day. McLeod was undeterred. He dived to attack the nearest and his observer shot it down out of control with a burst from close range.

On 14 January 1918, flying with Lieutenant Reginald Key as his observer, McLeod shot down an observation

Alan Arnett McLeod.

balloon over Bauvin, south east of Lille, pressing home his attack despite the attentions of three Albatros scouts which were protecting the balloon. McLeod had to pull away sharply from the blazing wreckage of the balloon and Key shot down an Albatros which was attempting to attack them from underneath, the enemy scout breaking up in mid-air. McLeod and Key fought off the two remaining Albatroses and returned. For this action McLeod was mentioned in despatches.

On 21 March 1918, the German forces launched a massive attack against the British 3rd and 5th Army fronts. By 25 March the 3rd Army front began to give way under the incessant attacks and the situation was desperate. The RFC was called upon to give every assistance to the hard-pressed ground forces and it concentrated more than a hundred machines on the 3rd Army front, both Army and Corps aeroplanes machine gunning and bombing the advancing enemy. However, no sooner was the crisis averted on the 3rd Army Front, than the brunt of the enemy attack switched to the 5th Army; on 27 March, the pilots and observers of the RFC were again fully engaged in ground strafing and bombing in an attempt to stem the German advance.

On the morning of 27 March, seven FK8s of 2 Squadron took off to bomb and strafe German positions at Bray-sur-Somme, near Albert. Visibility was extremely bad and Alan McLeod and his observer, Lieutenant Arthur Hammond MC, flying FK8 B.5773, were soon separated from their companions. Hampered by low cloud and fog, McLeod flew for two hours, trying without success to locate the target. Finally, low on fuel and feeling his way through the low cloud, McLeod arrived over 43 Squadron's aerodrome at Avesnes-le Comte. He decided to land and refuel, then make another attempt to find Bray-sur-Somme. The FK8 was heavily laden with bombs and the tailskid was broken in landing, but after lunch, with the FK8 refueled and the tailskid repaired by 43 Squadron's groundcrew, McLeod and Hammond took off again.

Weather conditions had not improved and it was two hours before McLeod managed to find his target. As he made to bomb an artillery gun emplacement, Hammond pointed out a Fokker Triplane below them. McLeod dived onto the Triplane and Hammond shot it down out of control to crash on the outskirts of Albert. The action had attracted the attention of eight Fokker Triplanes from the Richthofen *Jagdgeschwader* and they dived to attack the lone FK8 from all directions. After some fighting, several of the Triplanes broke off the action, possibly because they were getting in each other's way, but *Leutnant* Hans Kirschstein of *Jasta* 6 pressed home his attacks, coming up underneath and firing a long burst along the entire length of the British two-seater. The FK8 was damaged and Hammond was wounded twice. McLeod turned away from Kirschstein's fire, but another Triplane, coming in for a beam attack, wounded him in the leg and hit Hammond again. Despite his wounds, Hammond fired at this Triplane – which he claimed exploded – but Kirschstein attacked again, hitting the FK8's petrol tank. The FK8 burst into flames, which rapidly burnt away the flooring of the cockpits, McLeod's flying boots, and the lower part of his flying coat. With the instrument panel and joystick beginning to smoulder, McLeod climbed out of his burning wicker seat onto the lower wing. Holding the burning control stick in one hand, he put the FK8 into a slideslip in an attempt to fan the flames away from himself and Hammond, who, with no floor in his cockpit, had climbed out to lie along the fuselage, holding on to the Scarff ring of his Lewis gun. Kirschstein turned away, confident of his second victory.

As the burning FK8 sideslipped down it was attacked by another Triplane whose fire wounded them both again. According to eyewitness reports, Hammond shot this Triplane down into the German reserve trenches, although it is difficult to see how he could have done this as by now he was lying across the fuselage of the FK8. In total, Hammond claimed four Fokker Triplanes shot down in this action but these claims are not reflected in the German losses.

An FK8 of 2 Squadron RFC.

The blazing FK8 finally crashed in a shell crater in No-Man's Land. McLeod was wounded in five places, badly burnt, and in shock, but Hammond was unconscious. McLeod knew that the bombs were still on their racks and the aeroplane might explode at any moment. He pulled his observer from the wreckage and set off through the mud to the British front-line trenches, carrying Hammond on his back. Behind him, one of the FK8's bombs exploded, its shrapnel wounding him again. McLeod struggled on, under intense machine gun fire from the enemy trenches, was wounded yet again and finally collapsed a short way from the British trenches. This part of the line was held by South African troops and they pulled the badly wounded pilot and his observer to safety under a heavy barrage from the German guns. Sixty years later one of the South Africans recalled, 'We attended their wounds but could not safely get them away until dusk. Both were badly burnt and in a bad way.' McLeod was conscious, but Hammond was too badly injured to talk. Both 'smelt terribly of burnt flesh.'

It was five hours before the two airmen could be taken through the reserve trenches to the nearest casualty clearing station, three miles away. They were attended to there, then taken first to Amiens and then to hospital at Etaples. McLeod was later transferred to the Prince of Wales hospital in London where his life hung in the balance for two months, his father traveling from Canada to be at his son's bedside.

Arthur Hammond was awarded a Bar to his Military Cross for his part in the action on 27 January 1918. He recovered from his wounds but his shattered leg had to be amputated. After the war he emigrated to Winnipeg, the home state of his pilot.

The award of a Victoria Cross to Alan McLeod was gazetted on 1 May 1918, both for his part in the air combat and for bringing in his wounded observer under fire, and on 4 September 1918 the young Canadian – at 19 years of age the youngest VC for aerial combat in the First World War – supporting himself on two sticks and accompanied by his proud father, received his VC from His Majesty King George V at Buckingham Palace.

Immediately after the investiture, McLeod went home to Canada, arriving in Winnipeg on 30 September, on the way back to his home town of Stonewall. Here his health gradually recovered, but on 6 November 1918, still weak from his wounds, he died from the influenza epidemic which was then sweeping the world. Alan Arnett McLeod VC was buried on 9 November, two days before the Armistice.

Lieutenant Frank Hubert McNamara

A native of Rushworth, Victoria, born 4 April 1894, Frank Hubert McNamara was the only Australian to be awarded the Victoria Cross for air operations during the 1914–1918 war. A school teacher at the outbreak of war, he learned to fly at Point Cook and took his Royal Aero Club certificate No.2254 on 20 October 1915. McNamara was posted to 1 Squadron, Australian Flying Corps, as adjutant and went with the squadron to Egypt on 16 March 1916. The Squadron had no aeroplanes of its own; these were to be supplied by the Royal Flying Corps in Egypt. Many of the men had never seen an aeroplane and only a few of the officers had flown. Some officers, including McNamara, were sent to England for more advanced training and the remainder of the squadron was divided into two detachments and attached to two RFC squadrons for training, each officer and man in the Australian squadron being an understudy of their RFC counterparts. At the end of six weeks, 80% of the Squadron had reached a high state of efficiency and was concentrated at Heliopolis until 14 June 1916 when a Flight was sent to Suez. The Squadron was not to operate from one aerodrome as a unit until the middle of December, when it was flying from Mustabig aerodrome. While at Mustabig the Squadron began to receive more up-to-date aeroplanes, BE2es replacing their BE2cs. Bristol Scouts were also issued but found to be too slow and were returned; additional Martinsyde G.100s being added to squadron strength, two allocated to each Flight.

Frank Hubert McNamara.

In 1916 the war in Egypt was being fought on two main fronts: southern Egypt and, to the east of the Suez Canal, the Sinai desert, which was either in Turkish hands or a No-Man's Land. The RFC had two squadrons in Egypt, one on the Canal at Kantara and Ismailia, the other at Heliopolis to cover southern Egypt. Opposing them was one German squadron, mounting operations against the Suez Canal.

McNamara had returned to Egypt on 24 August 1916, and after a refresher course with 22 Reserve Squadron RFC was posted back to 1 Squadron AFC, now operating from Kantara.

By mid-December 1916, the Egyptian Expeditionary Force was assembling before the line of El Arish. The Turkish forces retreated before the attack had begun, but a large force of the enemy was seen to be concentrated at Magdhaba and on 22 December the allied troops moved out to attack the position. The two Australian squadrons – 1 and 14 – flew air support for the troops, bombing the Turkish positions at Magdhaba and the railway bridge at Irgeig. Magdhaba fell to the Australian troops on 23 December. 1 Squadron was moving up with the troops and by the middle of January was based at Kilo, to the west of Masaid.

Frank McNamara was fully involved in 1 Squadron's operations, taking part in reconnaissance, photography, and bombing, and on 18 February 1917, flying BE2e 7133, he bombed the German aerodrome at Ramleh. After the attack he dropped a note for the enemy, in French, suggesting that they move their dummy aerodrome outside Ramleh to a more concealed location.

The terrain over which the fighting was waged was extremely hazardous. The desert was dangerous and inhospitable, with huge sand hills heaped up by the scorching winds. It was a vast, lonely place, terrifying to all but the Arabs. Added to this was the danger of a forced landing. For the aircrew it was almost certain death, either from the desert or the hostile tribesmen or Turks, both of whom were renowned for their cruel treatment of prisoners. Because of this, attempts were always made to rescue any crew which forced landed in the desert and it was against this background that the action took place which led to the award of the Victoria Cross to Frank McNamara.

On 20 March 1917, two BE2cs flown by Captains D.W. Rutherford and Lieutenant Drummond left Kilo to bomb the railway at Junction station, a vital Turkish supply point which had been the object of concentrated attacks since the afternoon of 9 March. The two BEs, carrying bombs, were each escorted by a Martinsyde G.100. These were flown by Frank McNamara and Lt. L. Ellis, and armed with six modified howitzer shells with a delayed action fuse of 40 seconds.

On arrival at the target the BEs bombed first, followed by Ellis. As Ellis finished his bombing run he saw an enemy aeroplane and set off in pursuit. McNamara, coming in behind Ellis, in Martinsyde G.100 No.7486, dropped three of his bombs on the train attacked by Ellis, then two more on the railway line. The last bomb exploded as it left the Martinsyde's bomb rack and a piece of shrapnel damaged the underside of the Martinsyde and tore into McNamara's right buttock. McNamara dropped two smoke bombs as marker guides for further attacks by allied aeroplanes, then turned for base. As he did so he saw Rutherford's BE2c No.4497 on the ground. It had been hit by the severe anti-aircraft fire and forced to land. Turkish cavalry and troops were already converging on the stranded aeroplane.

Despite his wound and the danger of capture, McNamara landed close by and taxied towards Rutherford who was attempting to set fire to the BE. With McNamara shouting to him to hurry, Rutherford abandoned his attempt to destroy the BE, ran to the Martinsyde, and hauled himself up to lay on the engine cowling between the center section struts. McNamara began his takeoff run, but due to his wound he had lost all strength in his right leg and was unable to use the rudder to counteract the swing of the Martinsyde to the left. Hoping that he could get the Martinsyde airborne before the swing became unmanageable, McNamara opened his throttle wider, but the swing became more

Martinsyde G.100 7486. McNamara won his VC in this Martinsyde, which has an additional undercarriage strut.

pronounced, the undercarriage collapsed under the strain and the Martinsyde crashed to one side, breaking the propeller and tearing off both wings. There was now only one hope. Jumping from his cockpit, McNamara fired a pistol bullet into the fuel tank of the Martinsyde and set the escaping petrol alight with a Very light. He and Rutherford then ran towards Rutherford's BE, under fire from the rapidly advancing Turkish cavalry. Behind the two Australians the Martinsyde blew up as the remaining bomb detonated.

Reaching the BE, Rutherford and McNamara rapidly assessed the damage. A tire had been ripped off in the forced landing, center section wires were broken, and a longeron appeared to be cracked. But the BE was their only chance of escape. Rutherford ran to the nose of the BE to swing the propeller, while McNamara scrambled into the rear cockpit. But McNamara found that the spare drums for the Lewis gun had fallen into the bottom of the cockpit and were jamming the rudder bar controls. He worked frantically to free the drums then shouted to Rutherford to swing the propeller. The engine burst into life and McNamara, throwing himself into the seat, began to taxi the BE, Rutherford scrambling onto the wing, then into the front cockpit. McNamara turned the BE into the wind and began to take off across the sand, the Turkish cavalry, now rapidly coming up behind, firing at full gallop, hoping to overtake the BE before McNamara could get it into the air. The BE was nearly stopped three times by patches of soft sand, heart-stopping moments for McNamara and Rutherford, but it finally left the ground and McNamara climbed away and set course for Kilo, 70 miles away.

The pain of his wound was now severe and McNamara nearly fainted several times during the hour and 20 minutes return flight, but by sticking his head out into the slipstream he managed to stay conscious and land safely at Kilo. On landing it was found that three of the BE's bombs were still in their racks.

McNamara was put on a hospital train, taken back to a base camp and from there transferred to hospital, where he stayed for the next five months. While he was in hospital he was promoted to Captain and the award of a Victoria Cross was gazetted on 8 June 1917. The extent of his wound was such that he was invalided home to Australia in September and discharged from the AFC on 31 January 1918. Having recovered, he was reinstated in September 1918 and became a flying instructor. McNamara was presented with his Victoria Cross by His Royal Highness the Prince of Wales on 27 April 1920 at the Federal Government House in Melbourne. On the inauguration of the Royal Australian Air Force in March 1921, McNamara was awarded a permanent commission with the rank of Flight Lieutenant. He was an Air Commodore by 1939 and was awarded a OBE in the New Year's Honours List. McNamara served throughout the Second World War, was awarded a CBE in 1945, and promoted to Air Vice-Marshal. He retired in 1947 and stayed in England, to live in Buckinghamshire. On 2 November 1961, Frank Hubert McNamara VC, CBE, BA, died after a fall in his garden.

Sergeant Thomas Mottershead

Thomas Mottershead, the only non-commissioned officer serving in the Royal Flying Corps to be awarded a Victoria Cross, was born in Widnes, Lancashire, on 17 January 1892. By the outbreak of war in 1914 Mottershead was studying engineering and although he was a married man with a baby son he enlisted in the Royal Flying Corps on 10 August 1914. With his engineering background he was posted to the maintenance staff at the Central Flying School, promoted to corporal in September 1915, and to sergeant on 1 April 1916. A month later he applied and was accepted for pilot training.

Mottershead trained at Upavon, gained a 2nd Class Pilot's certificate within a month, followed by a 1st Class Pilot's certificate on 9 June 1916. A fellow pupil at the CFS was a Sergeant James McCudden, who had gained his Royal Aero Club certificate on 16 April 1916. Both McCudden and Mottershead were good pilots, due no doubt to their technical ability, and they were both held at the CFS as instructors until 4 July, when they were posted to France with two other NCO pilots, Sergeants Haxton and Pateman. Two of the four were to win the Victoria Cross and all were to give their lives for their country: Haxton in October 1916 while flying with 11 Squadron; Pateman in February 1917 with 15 Squadron; Mottershead in January 1916; and McCudden in July 1918.

Thomas Mottershead.

Mottershead was posted to 25 Squadron from the pilots' pool. Based at Auchel, the squadron was equipped with the FE2b and FE2d, a large two-seater pusher aeroplane. Tom Mottershead immediately showed that he too had that aggressive spirit which made the early pilots of the RFC so formidable. On his very first patrol over the lines he attacked a hostile battery from low level and destroyed it by bombing from 1,000 feet. On 22 September 1916, flying FE2b 6998, with 2Lt. C. Street as observer, he attacked the railway station at Samain, going down to 1500 feet to bomb an ammunition train. He then flew above another train while Street machine gunned its entire length. As they pulled away from this attack, the FE was attacked by a Fokker E-type monoplane. Mottershead out-manoeuvred the Fokker pilot and Street shot the monoplane down in flames. For this action and for acts of sustained courage over the period of his war flying to that date – which included landing on a German aerodrome and machine gunning the sheds – Mottershead was awarded the Distinguished Conduct Medal, gazetted on 14 November 1916.

Mottershead's next posting was to 20 Squadron, stationed at Clairmarias and also equipped with FE2bs and FE2ds. In the middle of December, after nearly six continuous months of war flying, Mottershead was given leave and he spent Christmas of 1916 at home with his wife and baby son.

Mottershead returned to France and 20 Squadron in early January 1917 and on the seventh of the month he was detailed with another FE to carry out a fighting patrol in the area of Ploegsteert Wood. Mottershead's usual FE gave trouble and he was forced to use another, FE2d A39. Taking off after the other FE, with Lt. W.E. Gower as his observer, Mottershead soon caught it up and they arrived over their patrol area at 10,000 feet. The two FEs immediately came under attack by two Albatros scouts of *Jasta* 8 and were separated. Mottershead and Gower were attacked by *Vizefeldwebel* Walter Göttsch, who closed in under the tail of the FE and fired an accurate burst into the petrol tank. The FE at once erupted into flames. Gower grabbed a fire extinguisher from his front cockpit and played it onto Mottershead in an attempt to keep the flames from his pilot, but the back of Mottershead's flying coat was already alight. Watched by hundreds of troops in the front-line trenches, Mottershead brought the burning FE back over the allied lines. Knowing that his observer would have little chance of surviving a crash in the devastated area of the front line, Mottershead flew on. The troops in the reserve line trenches saw him circling, looking for a suitable landing field. Finding one, Mottershead deliberately turned the furiously blazing FE into the

An FE2d ready for patrol.

wind, anxious to make as normal a landing as possible. As soon as the wheels touched the ground the undercarriage collapsed and the FE stood up on its nose, throwing Gower clear, but with the rear-mounted engine threatening to crush Mottershead, trapped in his burning cockpit. With the aid of the troops who had swarmed to the scene, Gower pulled his pilot from the wreckage of his cockpit. Despite being badly burnt on his back, hands, and legs, Mottershead was fully conscious, talkative, and even cheerful, glad to be on the ground alive. He was rushed to the nearest Casualty Clearing Station, then transferred to hospital. While in hospital he was visited by his friend from 25 Squadron, another Lancastrian pilot, Sidney Attwater. Attwater found his friend badly burned but remarkably cheerful, even getting out of bed to watch a German observation balloon going down in flames over the front. Scolded by a nurse, Mottershead got back into bed. The next day he was dead, having fought for his life for four days and nights.

Mottershead was buried with full military honours in Bailleul Cemetery on 13 January 1917, with the whole of 20 Squadron present. The Squadron's temporary commanding officer, Captain G.J. Mohney-Jones, later wrote to Lilian, Mottershead's wife. 'At the funeral, as we lifted the wreath and Union Jack from the coffin, we sorrowfully knew that we had lain to rest one of the bravest men who had ever fallen in war.'

On 12 February 1917, Thomas Mottershead was awarded a posthumous Victoria Cross and it was personally presented to Lilian Mottershead by King George V at an open air investiture in London's Hyde Park on 2 June 1917.

Widnes, Mottershead's home town, opened an appeal for Lilian and her baby son. Over £1,000 was raised, a considerable sum in those days, but for some unaccountable reason neither Lilian nor her son received any money. Nearly 50 years later the sum was discovered and used to endow a Mottershead Scholarship at Widnes Technical College.

Walter Göttsch, Mottershead's victor on that fateful January day, was wounded two days later by Captain R.M. Knowles of 20 Squadron. Recovering from his wound, Göttsch was again in combat with 20 Squadron's FEs on 29 June 1917 and yet again shot down and wounded. On 10 April 1918, Göttsch with 20 victories to his credit and now the *Staffelführer* of *Jasta* 19, finally met his end, the return fire from an RE8 observer of 52 Squadron fatally wounding him, sending his Fokker Dr.I 419/17 down to crash behind the British lines.

Captain Andrew Frederick Weatherby Beauchamp Proctor

Andrew Frederick Weatherby Beauchamp Proctor – 'Proccy' to his friends – was born on 4 September 1894, the second son of Mr. and Mrs. J.J. Proctor of Mossel Bay, Cape Province, South Africa. After his matriculation in 1912, Proctor studied engineering at the University of Cape Town, gaining a second year pass in 1914. Like many others of his generation, Proctor was anxious to play his part in the war and in October 1914 he was granted a year's leave from his studies and enlisted in the Duke of Edinburgh's Own Rifles. After active service in

Andrew Frederick Weatherby Beauchamp Proctor.

the campaigns in South West Africa, during which he was transferred to the South African Field Telegraph and Postal Corps, he returned to his studies, completing a third year.

Early in 1917, Proctor enlisted in the Royal Flying Corps as a Third Class Air Mechanic and sailed for England. On arrival he was first sent to No.6 Officer Cadet Battalion, then, on April 13, to the School of Military Aeronautics at Oxford. Passing out from Oxford in six weeks, Proctor was posted to 5 Reserve Squadron, stationed at Castle Bromwich. On 25 May, Proctor had his first flight, in Maurice Farman Longhorn No.7355, and four days later he repeated the experience; the combined total time for the two flights, of circuits and landings, was 45 minutes.

On 2 June 1917, Proctor was posted to 24 Reserve Squadron, commanded by Major B. McEwan, and over the next seven days, instructed by 2Lt. Pyecroft, he made 11 flights, a total of nearly four and a half hours in the air, and was judged ready to solo.

On 10 June, the future VC took off on his first solo flight. It was an inauspicious start, Proctor crashing Maurice Farman No.A6888 on landing, wiping off the undercarriage and lower port wing after a 25 minute flight. Undeterred by this unfortunate start, Proctor flew solo as much as he could in the following week and by 16 June he had flown a total of 9 hours 50 minutes, 4 hours 35 minutes of which were solo.

Within the next few days, Proctor was posted to 84 (Bombing) Squadron at Lilborne, continuing with his flying instruction and gaining valuable experience by flying both Sopwith 1½ Strutters and Avro 504s. On 29 July he was awarded his 'wings.' Ten days later Major W. Sholto Douglas took over command of 84 Squadron.

With the arrival of Sholto Douglas, 84 Squadron's role was changed to that of a fighter squadron and the change is reflected in Proctor's logbook, where he now began to enter flights in Nieuports, Sopwith Pups, and Camels. In late August the Squadron's first SE5as began to arrive, the type it was to fly in France, and on 20 September Proctor made his first flight – of 40 minutes – in an SE5a. The following day the Squadron flew to France and was based at Estrée Blanche, coming under the orders of 9th Wing.

Sholto Douglas – later Marshal of the Royal Air Force, Lord Douglas of Kirtleside – wrote of his impression of Proctor at this time. 'He was a wiry little man with black hair, a swarthy complexion, and snapping dark eyes, and from the beginning he showed that he was both courageous and a first-rate shot. He was so small (Proctor was just five' 1". Author) that we had to make special adjustments to the rudder bar and seat of his SE5.'[1]

Over the next two months, Proctor flew in the squadron patrols, taking part in 19 missions during the period 21 October to 22 November. On 11 November the Squadron had moved to a new base at Izel-le-Hameau, in the Third Army area, and it was flying from this aerodrome that Proctor had his first taste of success, helping Captain Edward Pennell to destroy one balloon and forcing another down. It was the first taste of what would become almost an obsession with the young South African: the destruction of enemy observation balloons.

On 5 December, testing his engine at 10,000 feet over Arras, Proctor saw an enemy two-seater 2,000 feet above him. He closed with the enemy machine over Cambrai, giving it a burst from both guns, but he underestimated the enemy gunner, who put accurate shots into the SE5a, hitting both the tailplane and control wires, before his pilot dived vertically away from their attacker. Proctor was not awarded the two-seater as a victory, but he entered it into his logbook as a probable.

On 1 January 1918, 84 Squadron moved again, this time to Flez, and it was from this aerodrome that Proctor scored his first victory. Taking off at 1:55p.m. on an offensive patrol, he and an American from Massachusetts, Lt. Jens Larsen, attacked a pair of enemy two-seaters three miles north east of St.Quentin, shooting them both down. That Proctor was still learning his trade is evidenced by his admission that he had fired over 400 rounds at the two-seater, from very close range.

After a fortnight's leave, Proctor returned to the squadron on 28 January, and four days later, on 2 February, again flying with Captain Pennell, he attacked

1. *Years of Combat*. Sholto Douglas. Collins. 1963.

Major William Sholto Douglas, CO of 84 Squadron in 1918.

an observation balloon, forcing the observer to take to his parachute and the balloon to be hauled down. Further success over enemy airplanes seemed to be eluding him, however, and it was not until 15 February, that he and Lt. Anthony McCudden, a younger brother of the 56 Squadron ace, James McCudden, attacked a two-seater, which they sent down out of control after killing the observer. This two-seater was awarded as Proctor's second victory, shared with McCudden. Two days later, on 17 February, Proctor shot an Albatros D.V down out of control for his third victory and this was followed by another Albatros on 19 February, which he destroyed south east of La Fere, again confirmed by McCudden. Nine days later a Pfalz D.III fell out of control to Proctor's guns, again in the area of La Fere.

Proctor was now fighting daily combats, but it was not until 15 March that he scored his next victory. Flying in a patrol led by Captain Kenneth Leask, the SEs stalked a DFW, which crossed the lines near Bellincourt. Leask dived to attack the enemy, followed by Proctor, but Leask lost his pressure in the dive and was forced to break off his attack. Proctor zoomed under the tail of the two-seater and opened fire, wounding the observer. Closing to within 100 yards, Proctor opened fire again. The two-seater turned over on its side and began to spin slowly down, the observer hanging over the side of his cockpit. Proctor followed the DFW down and saw it crash just south of Villeret, a mile over the British side of the lines. This DFW was from *Flieger Abteilung (A)* 207 and was crewed by *Gefreiter* Johann Ommen and *Leutnant* August Wagner; both were killed.

Two days later, during an escort to the bombers of 5 Naval Squadron, Proctor scored again, sending down two Albatros scouts out of control and destroying a Pfalz D.III which broke up in mid-air. Fighting with more Albatros scouts, one of them hit the radiator of Proctor's SE5a D259, deluging him with water. Proctor was four miles over the wrong side of the lines and he broke off the action and turned for the safety of the British trenches, where the pursuing enemy scouts left him. The engine of the SE5a finally seized solid over his aerodrome at Flez, but Proctor landed safely.

The German spring offensive opened on 21 March 1918 and over the following days 84 Squadron were fully engaged in ground strafing and bombing the rapidly advancing enemy troops. Proctor flew a patrol in the morning on the first day of the attack, bombing an ammunition dump at St.Quentin. Taking off again in the afternoon, he shot up and bombed troops concentrated near Houton Wood. He again flew two patrols the next day, but flew four on 24 March, then two or three daily

'Proccy' with friends in 84 Squadron. Left to right: Simpson, Highwood, Proctor, Boudwin (USAS), summer 1918.

until 2 April. With the British Army in full retreat, fighting desperate rearguard actions, 84 Squadron was forced to move its base three times over the next seven days to escape the advance of the German troops – to Champien, Vert Galant, and Conteville. Finally, the Squadron moved to Bertangles, where it was to stay until September.

Proctor left for ten days leave on 7 April. Before he left he learnt that he had been awarded a Military Cross for his work in France; when he rejoined the Squadron on 16 April he had been promoted to Captain and was given command of 'B' Flight.

The first few patrols as a Flight Commander brought Proctor no success, but on 23 April, after bombing enemy trenches, his Flight attacked nine Fokker Triplanes. Proctor selected the leader – flying a Triplane with green wings and a black tail – closed the range to 200 yards and opened fire. The Fokker dived, then went down like a leaf, completely out of control over Framerville, south of Bray, at 4:40p.m. For the remainder of April and into May, Proctor led his Flight on bombing and ground strafing missions and there were few opportunities to engage enemy aircraft. On 4 May, in an attempt to bring up the German fighters, 84 Squadron bombed the enemy aerodrome at Cappy, the base both of *Jagdgeschwader* I, commanded by Manfred von Richthofen's successor, *Hauptmann* Wilhelm Reinhard, and *Jasta* 5. A total of 52 bombs were dropped in the raid but there was no opposition.

On May 9, returning from a bombing raid in the vicinity of Mericourt and Cappy, Proctor saw two SE5s in combat with eight enemy machines and he took his Flight into the fight, shooting down an Albatros D.V out of control west of Villiers Bretonneux. The next day, while flying escort to the DH4s of 205 Squadron on a bombing raid to Rosières, Proctor caught a Rumpler south east of Bray and destroyed it at 2:10p.m. The little South African was now getting into his stride. The skills, often subconsciously learned, were now coming to fruition; May was to be a very successful month.

The rest of May brought Proctor another nine victories: four Albatros D.V scouts destroyed, with another three out of control and one captured. His most successful day was 19 May, with a two-seater down out of control over Vauvillers at 9:45 in the morning, followed by an Albatros D.V destroyed 15 minutes later over Weincourt. The evening saw another Albatros fall under his guns, at 6:35p.m., to be captured at Cachy. This was possibly *Oberleutnant* Hans Witt. Two other Albatros D.Vs destroyed in the same combat were shared with Lieutenants Macdonald, Oliver, and Biccard. The next day Proctor shot down a two-seater, which crashed near Hangard, but there was then a slight lull in his victories until 27 May, when he shot down another Albatros D.V out of control over Mézières. During a fight the next day over Villers Bretonneux, a shot hit Proctor's SE5a C1772, shattering the propeller. The excessive vibration broke two longerons but Proctor managed to force-land at Allonville. May 29 saw another Albatros D.V fall out of control south-east of Fricourt during an evening patrol, followed by another two days later near Hangest. During May, Proctor flew 63 hours 25 minutes of war flying.

The new month brought six more victories, including three balloons, the first of the 16 he was to destroy. The first of these victories came on the first day of the month, with an attack on two balloons south east of Fricourt. Proctor shot the first one down in flames and his attack on the second forced the observer to take to his parachute. A water pipe then burst, and Proctor had to return to base. A patrol on the evening of 3 June saw combat with a formation of mixed enemy fighters and a Pfalz D.III was forced down on the aerodrome at Bertangles. Proctor took the opportunity to make a 20-minute flight in the enemy fighter. On 5 June, Proctor took part in the destruction of a Rumpler, shot down 2,000 yards west of Moreuil, sharing the victory with Lt. William Southey, a fellow South African. The next day saw two more victories over balloons, the first north west of Proyart, the second further north, near Bray.

Proctor had now served nine months in France and on 7 June, Sholto Douglas wrote to the Officer Commanding 22nd Wing, Lt. Col. T.A.E .Cairnes, recommending an immediate award of a DSO to Proctor. Douglas ended the recommendation with 'He has now accounted for 26 enemy aircraft. He is a remarkably fine patrol leader and has on many occasions when leading large formations shown the greatest skill in aerial tactics. His keeness and courage are beyond praise.'

On 9 June, Proctor and Lts. Macdonald and Biccard attacked a two-seater, killing the observer before the pilot managed to evade them in the cloud cover. Two days later, however, Proctor had better luck, shooting down an Albatros D.V near Montdidier. Forty minutes later he destroyed a Pfalz D.III, shooting it down in flames to crash north of Cayeux. Proctor returned to his balloon attacks on 13 June, attacking the first east of Contoire and sending it down in flames. The second balloon he attacked was heavily defended and Proctor was forced to break off his attack by accurate machine gun fire from its defences, after killing or wounding the observer, who fell back into his basket.

Proctor went on leave on June 14, and was further rested by participating in a recruiting drive before returning to France and 84 Squadron on 8 August for his second tour of duty. His eye had not lost its skill and August 1918 was to be his most successful month, with 15 victories.

Proctor started his run of victories on 8 August, the first day of the Allied offensive. Orders were to destroy all enemy balloons, orders after Proctor's own heart, and flying with Sholto Douglas and 2Lt. Boudwin, he attacked two enemy balloons near Rosieres, destroying one. Proctor claimed two enemy aircraft as probables on 9 August, but although one was confirmed by a pilot of 32 Squadron, official credit was not given. There was no refusal on 11 August, however. Proctor attacked a Fokker D.VII, over Villers Carbonnel, and shot it down out of control at 6:10p.m. for the first of his victories over the new type.

On 14 August the DH4s of 205 Squadron were returning from a bombing raid, escorted by the SEs of 84 Squadron, when a strong force of 18 enemy fighters tried to attack them, closing in on their tails. The SEs thwarted the attacks, shepherded the bombers back to their base, and then returned to fight the enemy fighters, whose number had swelled to nearly 40. Aided by SE5as from 92 Squadron, the SEs of 84 Squadron pressed home their attack, Proctor shooting down in flames a Fokker D.VII with a red fuselage and sending down another out of control.

On 16 August, Proctor shared in the destruction of a DFW, which fell in flames over Estrees, the enemy observer jumping from his burning airplane. Five days later he shot down an Albatros two-seater over Fay for his 34th victory. On 22 August orders were again to attack the six balloons along the 3rd Corps front. Proctor approached the first, over Assevillers, and shot it down in flames. He then flew along the line of the five remaining balloons. The next, over Hem, failed to ignite, although Proctor emptied half a drum of Lewis into it. He had no better luck with the other four, although his attacks forced all the observers to jump. In the evening, no doubt still unhappy with his failure to shot down the balloon over Hem, Proctor returned. This time he made no mistake. The balloon went down in flames.

Proctor wearing his VC ribbon.

On 24 August, while patrolling over Brie, Proctor attacked a Fokker D.VII, sending it down out of control. Fifteen minutes later he saw a DH4 under attack from another Fokker D.VII. After a long burst the enemy fighter went down in a cloud of smoke and flames east of Brie. The following day, on patrol with Lts. N.H. Goudie and I.P. Corse, Proctor destroyed another Fokker D.VII, this time over Tempeux. Climbing for height, the patrol next attacked a Rumpler flying at 20,000 feet. The enemy two-seater burst into flames under the guns of the SEs and finally crashed near Vrely. At the end of the day, Sholto Douglas again recommended Proctor for a DSO, pointing out that since his last recommendation on 7 June, Proctor had increased his victories to 40, including seven balloons, and adding 'He has a wonderful power of seeing through enemy ruses and crooked tactics which has often stood his patrol in good stead.' The award was finally granted on 29 October.

On 27 August, it was balloons again, Proctor destroying two: one at 8:40a.m., the other over St.Quentin an hour later. On 28 August another Fokker D.VII fell under the South African's guns, shared with Lt. Falkenberg.

Not all Proctor's fighting was done in the upper air. The German Army was in full retreat and the fighter squadrons of Twenty-Second Wing were both harassing the German infantry and seeking out those strong points which were holding up the advance of the allied troops. On 4 September Proctor led his patrol across the line of the allied advance, flying further into enemy held territory, to Roisel, east of Peronne. Flying at 1,000 feet, Proctor could see very little movement on the roads and the anti-aircraft fire was both slight and inaccurate. Diving to within 50 feet of the ground, Proctor saw that there were few German troops in the area. Flying at only 20 feet, the SEs contour-chased south along the German lines to Hancourt and Mons-en-Chaussee, where they crossed the long straight road that runs east from Amiens. The trenches here were only lightly held and the British fighters flew on towards their old aerodrome at Flez, evacuated during the German attacks in March. A mile or so to the north of their old field, a German field gun was either foolish, or brave enough to open fire on them. The SEs quickly silenced it with low level attacks, killing the gun crew.

Proctor then led the patrol to the west, still flying very low. They soon saw the very thing they were looking for: a German machine-gun team was waiting in a sunken road, hoping to catch the advancing British troops. Proctor led the Flight into an attack and killed the machine gunners. The Flight then flew back across the advancing troops, who signaled that there were more German troops ahead of them. Proctor led his Flight back, flying only ten feet above the ground, and attacked the enemy trenches and machine gun nests, clearing the way for the following troops.

On 7 September, Proctor fired another balloon, one in a line from Cambrai to St.Quentin, the rest of his Flight destroying several others. On 15 September the Flight approached a line of five enemy balloons, guarded by nine Fokker D.VIIs. Proctor waited until the Fokkers were over the furthest balloon, then attacked the nearest, sending it down in flames over Bellicourt. The Fokkers made for the SEs, but they were too late. The Flight dived to ground level and contour-chased back over their own Lines.

Proctor had a lucky escape on 20 September. During a combat with Fokker D.VIIs, one got onto the tail of Proctor's SE5a and began shooting from close range. Luckily for Proctor, his predicament was seen by Lt. Falkenberg, who broke off his own attack to shot the offending Fokker off his Flight Commander's tail.

Proctor shot down another balloon on 24 September, with his Flight claiming six more. Proctor was now flying at least two patrols a day but his next victory was not until 27 September, when he destroyed another balloon, two miles to the west of Crevecoeur. The last day of September saw Proctor chased back to the lines by a superior force of enemy fighters and his SE5a was hit in the engine by machine gun fire from the ground.

On 1 October a full strength squadron patrol met a strong force of 30 Fokker D.VIIs escorting a trio of two-seaters. The Fokkers all sported different coloured tailplanes: yellow, red, blue, white, and black and white striped; the two-seaters all had orange noses. In the ensuing fight, Proctor shot down a blue-tailed Fokker, which crashed south east of Fontaine, then another – this one with a red tail – to crash in flames to the south east of Ramicourt.

The next morning Proctor was up early and destroyed a balloon over Selvigny at 8:30a.m. On 3 October a Squadron patrol met another strong force of 30 Fokker D.VIIs and seven Rumplers. Proctor claimed a Fokker out of control from the fight, going down over Mont d'Origny. An hour and 20 minutes later, at 11:45a.m., Proctor attacked a balloon over the same area and shot it down in flames. Two days later, on 5 October, Proctor destroyed his last balloon, west of Bohain.

The German Army was now in full retreat. On 8 October, 84 Squadron moved to Bouvincourt and on patrol from the new aerodrome later that day Proctor caught a Rumpler north east of Maretz and shot it down, the enemy machine bursting into flames as it hit the ground. Proctor followed the Rumpler down and in doing so was hit in the forearm by small arms fire from the ground. Despite his wound, he attacked a balloon on the way back to Bouvincourt, but this brought no positive result and Proctor returned. It was the end of his war.

Immediately after Proctor had been taken off flying and sent home, Sholto Douglas sat down and, 'with the help of a few stiff drinks to work myself into a sufficiently inspired mood,' he wrote a recommendation that Proctor should be awarded a Victoria Cross. The award was gazetted on 30 November 1918. It read, in part.

'In all he has proved himself conqueror of over 54 foes, destroying 22 enemy machines, 16 enemy kite balloons, and driving down 16 enemy aircraft completely out of control. Captain Beauchamp Proctor's work in attacking enemy troops on the ground and in reconnaissance during the withdrawal following the battle of St.Quentin from March 21st, 1918, and during the victorious advance of our armies commencing on August 8th, has been almost unsurpassed in its brilliancy, and as such has made an impression on those serving in his squadron and those around him that will not easily be forgotten.'

After Proctor was discharged from hospital in March 1919, he was selected to go to America in aid of the Liberty Loan Bond drive. The idea was that a 'flying

En route for America in 1919 on board the *Northland*. From left to right: Captain Beauchamp Proctor, VC, DSO, MC and Bar, DFC (54 victories); Major S.E. Parker; Major F. Holliday, DSC, MC (eight victories); Major Gerald Joseph Constable Maxwell MC, DFC (27 victories); Major P.F. Fullard, DSO, MC and Bar (46 victories). This historic photograph was taken by the remaining pilot of the group, Captain H.W. Woollett, DSO, MC and Bar, CC, L.d'H. (35 victories).

circus' should barnstorm across America, giving flying displays and mock aerial combats to encourage the population to buy Victory or Liberty Bonds. The pilots chosen for the tour included: Gerald Constable Maxwell (27 victories); Philip Fullard (46 victories); Henry Woollett (35 victories); Fred Holliday (17 victories); and Proctor. In addition to their impressive total of aerial victories – 179 enemy aircraft between them – the contingent held one Victoria Cross; four DSOs; five MCs with three Bars; two DFCs, a *Croix de Chevalier* and a *Légion d'Honneur*. With a Major Parker in overall command, the group embarked for the USA on 27 March 1919, sailing in the Northland, a troop transport taking Canadian troops home to Canada.

On arrival in America, the pilots were split into three groups, each group to tour a different region in special trains. One train toured the Eastern Seaboard States, another the Middlewest, and the third the far West and Pacific States. It was intended that every important city and town should be visited, with the pilots giving demonstrations and talks. The trains were elaborate affairs, with sleeping accommodation for 30 officers and 80 other ranks, two restaurant cars and additional trucks carrying 18 airplanes, spares, and workshops.

In addition to the British pilots, there were seven American service pilots in the tour, an Italian, and several aerobatic and specialist pilots. The airplanes used were a pair of Fokker D.VIIs – still carrying their German markings – a Nieuport 27; several 'Curtiss' SE5as; a 'Curtiss' SVA5 and a number of Curtiss Jennies. The tour was highly successful.

Proctor returned to England in July 1919 and was sent to the RAF College at Cranwell. In November he was granted a permanent commission as a Flight Lieutenant and attended an investiture at Buckingham Palace to receive his Victoria Cross, DSO, and DFC. Given extended leave, Proctor embarked for South Africa in February 1920. He returned to England the following winter and was posted to 24 Squadron. His movements from this date are uncertain, but it is known that during the winter of 1920/21 he was stationed at Upavon, where it is believed he was training for the Hendon Air Pageant of 1921.

On 21 July 1921, Proctor took off in a Sopwith Snipe to practice aerobatics. While attempting to roll off the top of a loop, he lost control. With very little height to recover, the Snipe crashed at Enford, a mile from Upavon, and Proctor was killed.

Anthony Frederick Weatherby Beauchamp Proctor, VC, DSO, MC and Bar, DFC, was buried with full military honours at Upavon.

Perhaps Proctor's old Commanding Officer in 84 Squadron, Sholto Douglas, should have the last words. 'For all his size, that little man had the guts of a lion.'

Major Lionel Wilmot Brabazon Rees

Lionel Wilmot Brabazon Rees was born in Castle Street, Carnarvon, Wales, on 31 July 1884, the son of a local solicitor. He grew into a sturdy, strong, and intelligent boy, if a rather solitary one, though this last was from choice rather than accident: being shy, he found it difficult to make friends, but those he did make were his for life. One such friend, recalling Rees' childhood, remembered that kite flying was popular when they were small boys. Rees' box kite, larger than those of the other boys, could soar the highest, and when Rees was given a camera for his birthday, he at once conceived the idea of attaching it to the kite in an attempt to take an aerial photograph of Carnarvon: a portent of things to come.

In 1885, Rees was sent to boarding school, to Eastbourne College in Surrey, and while there he decided on an army career. Rees was placed in the specialized Army Class to prepare him for the entrance exam to one of the military academies, and in 1901 he passed into the Royal Military Academy at Woolwich – the 'Shop.'

Rees had an inauspicious start at the RMA but by sheer hard work he passed out 16th of 83 candidates, and in addition won the Tombs Memorial Prize, awarded to the cadet with the highest marks in the finals. At some stage during his time at Woolwich, Rees decided to apply for the Royal Garrison Artillery, possibly because there were far greater opportunities for service overseas, and on 23 December 1903 he was commissioned as a 2nd Lieutenant in the Royal Garrison Gunners.

During following years Rees was stationed both in Gibraltar and Sierra Leone and it was while serving in Africa that he grew to love the tropics. In addition, while in West Africa, Rees grew to admire and respect the black troops under his command and, together with his brother officers, turned them into a highly efficient force. Again, this was a theme which would return in later life.

Despite his interest in the work in West Africa, Rees felt it had begun to lack the challenge and excitement he needed, and on his posting home to England in 1912 his interest was to turn to aviation and the newly-formed Royal Flying Corps. Determined to enter this new element, Rees took flying lessons at the Bristol Flying School at Larkhill, and on 7 January 1913 he was awarded Royal Aero Club Certificate No.392.

At the end of 1912 the War Office had let it be known that any officer who wished to fly with the RFC could be seconded to the Corps from his regiment on gaining his RAeC certificate. The stipulation was that the officer should learn to fly at his own expense – between £70 and £80 – which would be refunded if he was selected for the RFC. Although Rees had qualified as a pilot he did not go on to the Central Flying School to be trained by the RFC, and was again posted to West Africa, but on leave in England in the summer of 1914, war was declared and Rees immediately volunteered for the RFC. On 10 August 1914, he was posted to the Central Flying School for a five week course to train as a service pilot; he completed this on 12 September, was awarded certificate No.221 and posted to 7 Squadron.

On 11 October 1914, Rees was detailed to take two men, and a truck containing spares, to 6 Squadron, then stationed at Ostend, Belgium. This done, he was to then find a suitable location for an Aeroplane Park. He duly delivered the stores to 6 Squadron, but the German Army was rapidly advancing into Belgium and Rees was advised to forget his further orders and return to England as quickly as possible. However, the situation was chaotic and Rees stayed with the squadron until 22 October, when he returned to 7 Squadron in England. On his return, Rees was promoted to Captain and his new rank made him eligible to command a Flight; in February 1915 he was posted to Netheravon, to take command of 'A' Flight of 11 Squadron.

In France the war in the air had taken a more aggressive turn. The main role of the air forces of the combatants during 1914 had been one of reconnaissance, but by the end of the year pilots and observers were taking various weapons on their daily sorties and exchanging fire with enemy aeroplanes. It was obvious that a new type of squadron was needed, a squadron which could fight in the air, in addition to its reconnaissance and photography work. Consequently, 11 Squadron was formed on 14 February 1915, equipped with the Vickers FB5 – colloquially known as the 'Gunbus,' a two-seat pusher aeroplane, powered by a 100hp Gnome rotary engine and armed with a Lewis gun for the gunner/observer.

On his arrival at Netheravon, Rees found he was the Squadron's first officer, but he was soon joined by his fellow Flight Commanders: Captain 'Pip' Playfair, commanding 'B' Flight, and Captain C. Darley, commanding 'C' Flight. The Squadron's pilots were arriving from the flying schools and during the spring and early summer of 1915, were given further instruction by the resident instructors at Netheravon and by Rees and his fellow Flight Commanders. By July the Squadron was ready for active service and on the 27 July 1915, it left for Vert Galant in Northern France. Engine trouble dogged the move and it was not until 29 July that all the Squadron's aeroplanes were safely at their new base. One of the later arrivals was A.J. Insall, an observer, who later wrote: 'On arrival we found our senior Flight Commander, Captain L.W.B. Rees, covered in oil and as happy as Larry, working like a navy on his machine, which had already collected a sprinkling of scars from a scrap with an Aviatik or LVG two-seater the day before.'

The role of 11 Squadron was to fly 'aggressively defensive patrols' over the 3rd Army Front and to photograph the German back areas and front line and reserve trenches. Rees threw himself into the work with all his customary enthusiasm, his cardinal rule being that the enemy should be attacked whenever he was seen: "...no matter where he is, be he alone or accompanied, go for him and shoot him down.' With his commanding presence and possessing, as one officer put it, 'the most captivating eye-twinkle of any officer I can recall,' Rees was an inspiration to the entire squadron.

Typical of Rees' style of fighting was his second combat. On 31 August he attacked an LVG at 7,000 feet near Bapaume, and the German observer returned fire with an automatic rifle before his pilot dived out of range. The LVG then climbed back, fired off a few shots at the Vickers, then retreated again, keeping up these tactics for nearly 20 minutes until Flight Sergeant Hargreaves, Rees' gunner, had exhausted all four drums of his Lewis gun ammunition. Rees was furious. He broke off the action and flew back to Vert Galant. As he landed he could be heard above the noise of the Gnome engine, which he kept running, shouting for more ammunition. Rees took off again and slowly climbed back to the scene of the recent fight. The LVG was still there, carrying out its work a thousand feet below the Vickers. Rees attacked and Hargreaves fired a complete drum at the two-seater, which immediately dived for the cloud cover, followed by the Vickers. The steepness of the Vickers' dive caused the engine to stop; unable to follow the enemy machine, Rees turned for home. Another Vickers from the Squadron, crewed by Lt. H.A. Cooper and A.J. Insall, saw the LVG break out of the cloud and fall in a series of spirals before they lost sight of it. Reports were later received from the ground forces to say an enemy machine had been seen to crash.

Rees also carried out the more mundane photo-reconnaissance work of the Squadron with his customary zeal. A series of photographs, taken by his observer Hargreaves, earned high praise from 3rd Army Headquarters, which described them as 'the finest series of photographs ever taken in France to that date.' During this patrol, Rees and Hargreaves attacked an Ago, sending it down out of control, and it was later seen to crash while attempting to land. Rees was awarded the Military Cross for this action, while Flight Sergeant Hargreaves won a Distinguished Conduct Medal.

Nine days later Rees and Hargreaves were in action again, attacking an Albatros two-seater near Gommecourt. Hargreaves emptied a full drum of Lewis into the enemy machine and the pilot dived for the cover of a cloud, followed closely by the Vickers. Hargreaves fired another drum and the Albatros went into a spiral dive. The British anti-aircraft guns now opened up and hit the starboard wing of the Albatros, which broke off at 5,000 feet. The wreckage hit the ground behind the British lines and it was found that the pilot had been hit by a bullet in the head and was dead before the guns had opened fire. The Albatros was awarded to Rees and Hargreaves.

On 31 October, Rees flew his last combat with 11 Squadron. A few days later he was posted to Home Establishment, to the Central Flying School as an instructor, and on 28 November 1915, was promoted to Major. Rees' rank now enabled him to be appointed as a squadron commander and on 1 February 1916 he was given command of 32 Squadron, working up to strength at Netheravon.

By June 1916 the RFC had four squadrons in France which could be termed true fighter squadrons, but only three of these, 24, 29, and 32, were equipped with a single type of fighter aeroplane – the DH2 – the fourth, 60 Squadron, having a motley assortment of Moranes. The DH2 was a single-seat pusher biplane, powered by the unreliable 100hp Gnome rotary engine, and armed with a single Lewis gun. Despite its doubtful engine and, at first, its unsavoury reputation for spinning – then scarcely understood – the DH2 was a compact little aeroplane and, although far from fast, was highly manoeuvrable: in October 1916, three DH2s from 29 Squadron held their own in a long combat with six of the new Albatros D.I fighters – a vastly superior aeroplane – and returned safely.

Lionel Wilmot Brabazon Rees.

On his arrival at Netheravon, Rees began the selection and training of his pilots and groundcrew. One of his pilots was a newly qualified 2nd Lieutenant G.H. Lewis. Lewis was impressed with his commanding officer. In a letter home he wrote that having Rees in command of the squadron made it 'mostly ideal' and told how Rees was not only technically expert on both engines and rigging and thoroughly knew the job, but was 'above all

The DH2s of 32 Squadron lined up on the aerodrome at Vert Galant in the summer of 1916.

a perfect gentleman.' In the same letter home, Lewis showed remarkable prescience, confiding. 'I shouldn't be surprised if he comes home with a VC.'

On 28 May, 32 Squadron left for France and was stationed at St.Omer for five days before moving on to St.Auchel on 4 June, sharing the aerodrome with 25 Squadron. On 7 June, the Squadron moved again, to Treizennes, where it came under the orders of 10th Wing, 1st Brigade, and flew in support of the 1st and 3rd Armies.

In addition to its DH2s, 32 Squadron had two Vickers ES1 scouts on charge: Nos.5127 and 7758. These little tractor scouts, known as Bullets, were also powered by the 100hp Gnome Monosoupape that powered the DH2. Vickers ES1 No.5127 appears to have been unarmed and was returned to the depot sometime during the first week of June, but 7758, armed with a Vickers gun, was flown by Rees during his Squadron's first operations in France.

In late June 1916 the British Army was about to launch the Battle of the Somme. On 24 June the Allied artillery opened fire on the German positions in the most intensive artillery bombardment of the war to date, the aim being to destroy the enemy front-line trenches and barbed wire defenses before the infantry assault. The squadrons of the RFC were given the task of insuring that no enemy reconnaissance machines crossed the lines to observe the preparations for the attack.

On 1 July, the first day of the Battle of the Somme, a patrol of DH2s from 32 Squadron left the ground at 3:40a.m., on escort duty for the BE2ds of 2 Squadron and FE2bs of 25 Squadron, which were to carry out a bomb raid on the railway station at Don. At 5:55a.m., Lionel Rees flying DH2 No.6015 and accompanied by a Canadian, J.C. Simpson, in DH2 No.7856, took off from Vert Galant. Simpson was to patrol the area La Bassée, Loos, and Souchez, while Rees patrolled the front line to await the return of the DH2s and the bombers. On the other side of the lines, ten German machines, Rumplers and Albatros two-seaters of *Kagohl* 3, commanded by *Leutnant* Erich Zimmerman, had taken off to patrol in the vicinity of Lens. As the German force crossed the lines, Simpson saw them and without hesitation attacked at once, despite the odds against him. Three of the enemy machines concentrated on Simpson and shot him down, with eight bullets in the head. His DH2 crashed and the wreckage was shelled by the German artillery.

At 6:15am, a few minutes after Simpson was shot down, Rees also saw the enemy formation. At first he mistook them for British machines, some of the bombers coming home, but as he approached Annequin one of the machines, marked with a large numeral 3 and a small cross on the fuselage, left the formation and fired at the DH2. Rees waited until the enemy machine was in range, then fired a drum of Lewis at it. His shooting was accurate: '...after the 30th round I saw the top of the fuselage splinter between the pilot and observer.' This machine turned round and went home. Rees then went on to attack a second machine, whose observer fired red signal lights calling for support from his escort. These, three in number, dived on Rees, but overshot and Rees closed in to a favourable position and fired a full drum of Lewis into the enemy two-seater, hitting it in the engine. A large cloud of blue smoke came out of two-seater's nose and it dived away for the safety of its own lines.

Rees next attacked five of the enemy machines, firing a Lewis drum at long range and dispersing the formation '...in all directions.' Rees saw the leader of the formation with two of his companions flying west, and he overhauled the lowest machine, opening fire at long range. The enemy turned left and dropped a bomb. Rees closed the range, but just as he was about to open fire, a bullet from the observer's gun hit him in the upper leg, putting it out of action. Rees fired one drum, but he had no rudder control over the DH2 and had to attempt to control his aeroplane by the joystick alone. At ten yards range, Rees stopped firing: he could see the enemy observer sitting in his cockpit but firing into the air instead of at the DH2. Having expended his drum, Rees drew his pistol, but in his haste he dropped it and it slid to the front of his nacelle. Feeling in his leg now returned and Rees found that he had some control over its movements. Looking round he saw the leader of the enemy formation heading for his own lines. Rees set off in pursuit, the German observer firing from long range at his pursuer. As the enemy machine reached the lines, two thousand feet above the DH2, which was at 9,000 feet, Rees pulled down his Lewis gun on its mounting and opened fire. 'I was using the Beliene sight fixed to the gun, but as the sun had only just risen it was not shining on the cross wires. Even without the cross wires, the tracer appeared to be going very near the target, simply through looking through the tube which aligned on the axis of the gun.' Rees fired a full drum of Lewis at the fleeing leader before turning away; it was impossible to climb up to his opponent's height and close the range further.

The German machines were now all dispersed and had returned to their own side of the lines. Satisfied with this and with his leg now becoming painful, Rees turned for home, making a good landing at Treizennes. After taxiing to the sheds Rees climbed painfully from the cockpit and sat on the grass to await the arrival of a tender to take him to the casualty clearing station. It was 6:50a.m.

Lewis was there when Rees returned and he later wrote to his parents: 'I told you he was the bravest man in the world...He landed in the usual manner – taxied in. They got the steps for him to get out of his machine. He got out and sat on the grass and calmly told the fellows to bring him a tender to take him to hospital.'

At the time of the combat it was only known that Rees had single-handed fought a German formation out of the sky, but the records of *Kaghol* 3 show that it lost two machines in the fight. One of these was the first machine which Rees attacked and sent down emitting blue smoke; the observer in the next machine attacked by Rees, seen by him to be sitting in his cockpit and firing into the air, was the *Staffelführer, Leutnant* Zimmermann. Zimmermann was killed by Rees' fire, and his pilot, *Leutnant* Wendler, was wounded. Their machine crashed near the little hamlet of Petit Hantay, near La Bassée.

Rees was awarded a Victoria Cross for his action on 1 July 1916; the award was gazetted on 5 August 1916, and on 14 December 1916 Rees was presented with his Cross by King George V.

The wound to Rees' leg was more serious than at first thought. The bullet had entered the lower part of his thigh and passed very close to the main artery. It had also caused some damage to the knee and thereafter Rees always walked with a slight limp. Rees hoped for a speedy recovery, but four months after the action, he was still in hospital and it was not until the end of 1916 that he was able to leave; even then he could only walk with the aid of two sticks, and he spent a further period of convalescence at the home of an aunt. By March 1917 he still could not walk any distance without the aid of his sticks and could not fully bend his injured leg.

In April 1917 America declared war on Germany and a British military mission was sent to the USA. Led by the British Foreign Secretary and former Prime Minister, Lord Balfour, the mission was known as the Balfour Mission; the officer chosen to advise on aviation was Major Lionel Rees. Rees returned to England in January 1918. He had been promoted to Lieutenant-Colonel in May 1917, but his leg wound precluded a command on active service in France. On 1 March 1918 Rees was given command of the No.1 School of Aerial Fighting and it was there that he saw out the remainder of the war.

Rees was awarded an Air Force Cross on 2 November 1918 and later given an OBE. He remained in the Royal Air Force until the end of July 1931, when he retired. Rees was restless in civilian life. He had bought a 34' 6" ketch, *May,* on his retirement from the RAF and on 2 July 1933 he set sail for the Bahamas, completing his single-handed voyage across the Atlantic by 21 October, when he anchored in Nassau.

Rees returned from the Bahamas to rejoin the RAF during WW2, relinquishing his retirement rank of Group Captain to serve as a Wing Commander. After serving in the Middle East and commanding RAF Helwan in Egypt, Rees reached retirement age in 1942 and he finally left the Royal Air Force in November, with the rank of Group Captain. Rees at once returned to the Bahamas, were he lived for the rest of his life, marrying a young local girl in 1947.

Rees became ill with leukemia in the early 1950s and returned to England for treatment. The treatment was not a success. Rees knew that he was dying and determined to return to his family in the Bahamas, where he was admitted to the Princess Margaret Hospital in Nassau in May 1955. Lionel Wilmot Brabazon Rees VC, OBE, MC, AFC died there on 28 September 1955 and was buried with full military honours in the RAF Cemetery.

Lieutenant William Barnard Rhodes-Moorhouse

Rhodes-Moorhouse has the distinction of being the first airman to win the Victoria Cross in the First World War. A Londoner, he was born on 26th September 1887, but his background was anything but usual, one of his grandparents being the daughter of a Maori chief.

Rhodes-Moorhouse was a slight but healthy boy, very self-reliant, full of vitality, and with a great love of speed. In his teens and early twenties he raced both motor cars and motor cycles, but by 1909 his enthusiasm was fired by the new element, the air, and he learnt to fly. By the time he took his Royal Aero Club Certificate No.147, on 17 October 1911 at Hendon, he had already been flying for two years.

In 1911 he traveled to America and, flying a Blériot powered by a 50hp Gnome engine, he took part in many of the cross country races then so popular in the United States. He won several thousand dollars in these races and won the Harbour Prize in San Francisco. Returning to England, he was in the forefront of the pioneers of the day, flying in exhibitions and flying meetings. In June 1912 he came third in the first aerial derby at Hendon. A month later, with his new wife, Linda, and a newspaper reporter as passengers, he took off from Douai in a Breguet to fly across the English Channel. The Breguet crashed on landing in Kent, England, but nobody was hurt and Rhodes-Moorhouse became the the first pilot to fly the crossing carrying two passemgers.

On August 24 1914, Rhodes-Moorhouse enlisted in the Royal Flying Corps as a 2nd Lieutenant and on March 20 1915 he was posted to 2 Squadron, based at Merville in France. Flying a variety of BE2 aeroplanes, 2 Squadron was engaged in reconnaissance and bombing duties and on 24 April 1915 the squadron was ordered by 1st Wing to bomb Roubaix, Tourcoing, and the railway

Lieutenant William Barnard Rhodes-Moorhouse.

This BE2b of 4 Squadron at St.Omer is typical of the type flown by Rhodes-Moorhouse.

junction and marshaling yards on the outskirts of Courtrai. Two days earlier, on 22 April, the Germans had launched a gas attack to begin the First Battle of Ypres, their forces had broken through at St.Julien, and they were bringing up reserves to exploit the breakthrough. Agents had reported that reserves were being concentrated around Ghent and were moving towards the Salient. RFC squadrons were ordered to attack trains and railway stations on the feeder lines.

At three minutes past three on the afternoon of 24 April 1915, Rhodes-Moorhouse took off from Merville in BE2b No.687, with orders to bomb Courtrai, some 35 miles behind enemy lines. In order to save weight, Rhodes-Moorhouse was flying without his usual observer, Sholto Douglas. The BE2b carried a single 100lb bomb. Rhodes-Moorhouse was fully aware that a lone 100lb bomb would do very little damage unless very accurately placed and, with no proper bombsight, the only way he could achieve this accuracy would be to bomb from a very low height. He knew that the chance of interception from enemy aeroplanes was slight, but that the ground defences around Courtrai station would be formidable. Nevertheless, reaching his target, he came down to 300 feet and dropped the bomb on the railway line, just to the west of Courtrai station. Rocked by the blast of the explosion, the BE was hit in wings and rear fuselage by shrapnel from the bomb and as Rhodes-Moorhouse struggled to regain control he came under attack, both from the machine-gun nests ringing the junction and rifle fire from the hundreds of troops who were assembling. A machine gun in the belfry of Courtrai church, firing from close range, raked the BE's fuselage, wounding Rhodes-Moorhouse in the thigh. In great pain and bleeding profusely, he regained control and dived to 100 feet in order to gain speed for a climb to safety but he was hit again, this time in the abdomen and hand. In spite of the pain and shock, Rhodes-Moorhouse was determined to return and make his report. Resisting the idea of landing and seeking immediate attention for his wounds, and fighting pain and constant dizziness, he flew on for 35 minutes, back to his base at Merville, landing at 4:15p.m.

Rhodes-Moorhouse was gently lifted from his shattered cockpit. The BE was severely damaged, with scores of holes and gashes from bullets and shrapnel. The cockpit was awash with his blood, but Rhodes-Moorhouse insisted on making his report before being taken to a casualty clearing station. His observer, Sholto Douglas, a future Marshal of the Royal Air Force, later wrote, 'When we helped him out of his aircraft and saw the full extent of his wounds we were amazed that the poor chap had ever managed to get back.' Viewing the bullet holes around the cockpit he would have occupied, Sholto Douglas realized, 'I would not have survived if I had been with him.'

Rhodes-Moorhouse made his report in General Trenchard's hearing and was then taken to a casualty clearing station. Despite every care, it was obvious that he was dying, and just before 2:30p.m., on 27 April 1915, with his great friend Maurice Blake and Padre Chavasse at his bedside, Rhodes-Moorhouse died of his wounds. In accordance with his last wish his body was taken back to England and he was buried on the family estate in Dorset. On 22nd May, 1915, the London Gazette carried the notice of the award of a posthumous Victoria Cross to William Barnard Rhodes-Moorhouse.

There is a postscript to the Rhodes-Moorhouse story.

Rhodes-Moorhouse on a stretcher waiting to be taken to a casualty clearing station.

On September 6 1940, Flying Officer William Rhodes-Moorhouse DFC, named after his father, was shot down and killed in his Hurricane during the Battle of Britain. He had joined the Royal Air Force in 1937 and during the battle of France had flown from Merville, the scene of his father's gallant action 25 years before. His ashes were buried beside the grave of his father.

Captain William Leefe Robinson

The first attacks by Zeppelins on the British Isles took place on 18 January 1915, when two Zeppelins, *L3* and *L4,* set out from their base at Fuhlsbüttel with orders to bomb military bases on the River Humber. Weather conditions were bad, but *L3* crossed the British coast over Norfolk at ten minutes to eight, the first Zeppelin to bomb England. Its companion airship, *L4,* crossed the coast 40 minutes later. Both Zeppelins returned safely to their base. Their attacks had killed a woman, a man, a small boy, and destroyed property. The Zeppelin terror had begun.

Nocturnal raids by the Zeppelins continued throughout 1915; by the end of the year the raids had killed 208 civilians and injured a further 432. Much property had also been destroyed or damaged. The war was now brought home to the war-weary British civilian population and it was angry at the ineffectiveness of the Home Defence forces, which seemed powerless to stop the Zeppelins.

The attacks resumed in January 1916 and continued throughout the spring and summer. As autumn approached it seemed that nothing could stop the raiders. Those aeroplanes which managed to find the giants in the night sky had failed in their attacks, the Zeppelins always being able to either climb rapidly out of range, or evade their attackers in the darkness. But this mood of helplessness was to suddenly change: the catalyst a young officer in the Royal Flying Corps.

William Leefe Robinson was born in India on 14 July 1895. At six years of age, William, and his elder brother Harold, came back to England for their early schooling, returning to India in 1903 and attending the Bishop Cotton School in Bangalore.

The Robinson boys again returned to England in 1907 and continued their education, enrolling in St.Bees School, Cumberland, in 1909, a school which would number no less than three Victoria Cross winners amongst its old boys. The Robinson boys were popular at St.Bees, especially the extrovert William. He was a mischievous boy, full of fun and practical jokes, and his sense of humour, throughout his tragically short life, endeared him to all who came into contact with him.

By early 1913 William was determined to try for Sandhurst, to enable him to join the Royal Engineers, and on August 14 1914 he entered the Royal Military College. On passing out, he was commissioned in the Worcester Regiment on 16 December 1914 and posted to Cornwall. Disappointed at not being able to see any action, he applied for a transfer to the Royal Flying Corps and on 29 March he was posted to France as an observer with 4 Squadron, stationed at St.Pol. On 8 May 1915, as observer to Captain E.F. Unwin, flying a Voisin, he was wounded in the arm by shrapnel from a bursting anti-aircraft shell. Taken to No.7 Hospital, the bullet and shrapnel were removed from his arm and he was given a month's leave in England.

On recovering from his wound, Robinson applied for pilot training, was accepted, and reported to Farnborough on 29 June, making his first flight under instruction the next day. After soloing on 18 July, Robinson took his RAeC certificate, No.1475, in a Maurice Farman on 28 July, and was posted to the Central Flying School at Upavon for advanced training on 14 August.

Robinson gained his 'wings' on 15 September and was posted to 19 Squadron, stationed at Castle Bromwich, but by 24 December he was at Joyce Green, seconded to 10 Reserve Squadron of the London Defence Force. This posting, a portent of things to come, only lasted a short while and 12 January saw Robinson back at Castle Bromwich. But it was a short-lived posting; on 2 February he arrived at Suttons Farm, near Hornchurch in Essex, to join the strength of 19 Reserve Squadron, commanded by Major T.C.R. Higgins.

In March 1916, 18th Wing was formed to bring together all the air defence attachments for the defence of London, and on 15 April these became 39 Home Defense Squadron. The Squadron consisted of three

William Leefe Robinson.

Flights, each equipped with six BE2c aeroplanes: 'A' Flight was stationed at North Weald Bassett; 'B' Flight, commanded by Robinson, at Suttons Farm; and 'C' Flight at Hainault Farm, both in Essex, on the eastern outskirts of London.

On 31 January 1916, 15 home defence aeroplanes had taken off to intercept the Zeppelins; on returning, without having gained any success, 11 had crashed on landing and four pilots had been killed. During the spring evenings the Flights now all gained much-needed experience in night flying and by early April night takeoffs and landings were being carried out without mishap.

On the 25 April, Robinson had his first chance to combat a Zeppelin.

Hauptmann Erich Linnarz, commanding *LZ97,* arrived over the River Blackwater and changed course for London. Mistaking the River Roding for the Thames, Linnarz spread a line of 47 incendiary bombs between Fyfield and Ongar. Flying on through the night, *LZ97* dropped its next bomb on Fairlop railway station – very near the base of the 39 HD Squadron – damaging seven small cottages. Other bombs fell harmlessly into the open fields of the area. The next bomb was dropped over Newbury Park, before the Zeppelin arrived over Ilford. Flying west from Ilford, *LZ97* came under heavy and accurate anti-aircraft fire over Seven Kings, where it came under attack from Robinson and Lieutenant Arthur Harris.

Robinson had taken off from Suttons Farm at 10:45p.m. Climbing to 7,000 feet he saw a great number of searchlights pointing north, indicating the position of the raider. Robinson saw the Zeppelin and immediately turned towards it, climbing hard. At just over 8,000 feet, directly under the huge target, he opened fire with his Lewis gun, but it had no effect on *LZ97,* 2,000 feet above the relatively tiny BE. Robinson opened fire three times but had five gun jams, could fire only 20 rounds, and his fire had no effect on the Zeppelin. *LZ97* made off to the north-east, followed by Robinson, but he lost it in the darkness and returned to Suttons Farm. Harris had also attacked the Zeppelin, but he too was frustrated by gun jams.

Linnarz and his crew made good their escape from the London area, shaken by the improved defenses of the capital. Over Belgium, and nearly home, they were attacked by two RNAS aeroplanes. One attacker was forced to land in the sea by return fire from the Zeppelin – possibly Flight Sub-Lieutenant S.G. Beare in Sopwith Baby 8140, who was rescued and interned in Holland – and the other made off. *LZ97* finally landed at its base nearly 12 hours after its takeoff.

Robinson was extremely angry and frustrated at his failure to down the raider and vowed that the next time it would either be him or the 'Zepp.'

Although the Zeppelin raids continued through the spring and summer, they were against Scotland, the Midlands, and Eastern counties of England, with only two raids over the Southern counties, and it was not until the night of 24/25 August that Zeppelin *L31,* commanded by *Kapitäinleutnant* Heinrich Mathy, arrived over central London, the first to do so since October 1915. This raid against the capital, which only Mathy reached, was largely abortive: 13 airships had set out but only four penetrated inland, five turning back over the sea because of the strong winds. Although the four which crossed the English coast headed inland and dropped their bombs, no damage was caused, other than that by Mathy, who dropped a total of eight incendiary and 35 high explosive bombs, killing nine people and injuring a further 40.

The next raid mounted by the airships took place on the night of 2 September 1916, when 16 airships set out for England: the largest concentration of raiders yet sent, with the German Army and Naval Airship divisions joining forces for the first time. It was to be a disastrous night for the airships.

The Schütte Lanz airship *SL11* was newly-commissioned in the German Naval Airship Division, flying to its base at Spich on 12 August after its initial flight tests. On the night of 2 September, commanded by *Hauptmann* Wilhelm Schramm, *SL11* arrived over the River Crouch at 20 minutes to 11. The flight to England had not been easy: severe weather conditions of extremely turbulent winds and heavy rain squalls, with several of the airships icing up over the sea, had caused some to abort their attacks and return. Schramm had pressed on and steered a wide circle from the River Crouch so as to approach the British capital from the north. Flying over Royston in Cambridgeshire, then

Robinson in BE2c 2693 at Suttons Farm. The groundcrew are displaying the center section of the top wing, which was damaged by Robinson's gun in his attack on *SL11.*

Hitchin in Hertfordshire, while Schramm attempted to fix his position, the airship arrived over London Colney, north of the capital, at about 1:00a.m. *SL11* was now nearing the London guns and Schramm lightened his ship by releasing high explosive and incendiary bombs, which fell into fields, doing no damage. Now considerably lightened, *SL11* climbed steadily for the protection of height from the defences of London. Forty-five minutes later, Schramm released another bomb and some incendiaries. The bomb fell on a stud farm at Enfield, killing three valuable racehorses. The raider flew inexorably on, dropping another bomb at Potters Bar in Middlesex, which ruptured a water main. Over Edmonton, a northern district of London, Schramm dropped more incendiaries. These did no damage and over the Alexandra Palace in Muswell Hill the airship was finally caught in the searchlights operating in Victoria Park and Finsbury. The guns immediately opened fire and Schramm was forced to turn and climb away. Over the districts of Ponders End, Walthamstow and Enfield, he dropped his final bombs, which damaged another water main and various properties, but caused no casualties among the civilian population. It was here that *SL11* was attacked by Robinson.

The War Office had telephoned Suttons Farm a little after 11:00p.m. with orders to intercept the approaching Zeppelins. 'B' Flight was on readiness. The night was very damp and foggy, but Robinson took off down the primitive pathway flares into the thickening fog, confident that conditions would be clearer at height. Robinson's BE2c, believed to be No.2693, was armed with a single Lewis gun using a new type of ammunition, explosive and incendiary bullets, and his orders were to patrol the area Suttons Farm to Joyce Green.

At first it seemed to Robinson that it was going to be another frustrating night. He was nearing the end of his patrol line when he sighted an airship, held in searchlights to the southeast. It was the *LZ98,* which had bombed Gravesend at 1:15a.m. At 12,900 feet, Robinson had a 800 feet height advantage over the Zeppelin and he was rapidly gaining on it when it suddenly vanished into cloud. After 15 minutes of fruitless search, Robinson abandoned the hunt. He had flown over his allotted patrol time and he made for Sutton's Farm, but as he started to lose height his attention was caught by a glow in the sky to the north: it was *SL11,* held in the searchlights over the northern outskirts of the capital. Robinson opened his throttle and climbed to attack the raider.

When he reached the airship, Robinson was several hundred feet higher and with this advantage he dived to 800 feet below it and made a head-on attack, firing a complete drum of Brock and Pomeroy bullets into the entire length of the huge hulk above him. *SL11* flew on, apparently undamaged. Robinson changed the Lewis gun drum and attacked again, this time firing along the airship's side. *SL11* flew steadily on. There was no return fire; the raider seemed oblivious of the attacking BE and took no evasive action.

Robinson was now determined to down the airship. He reloaded, and climbed until he was less than 500 feet below the raider, its bulk blotting out the stars in the night sky. Coming in from the rear, Robinson this time concentrated an entire drum of ammunition on one spot, just forward of the elevators and rudder of the airship. At first it seemed that his fire had again had no effect, but as his Lewis gun finally stopped, the drum exhausted, a dull red glow began inside the rear of the hull and within seconds the entire bulk of the airship was ablaze.

Robinson had to manoeuvre violently to avoid the burning mass, which threatened to engulf him as it fell. As he watched the stricken *SL11* fall, half dazed at the result of his attack, the flames lighting up his aeroplane 'as in the firelight,' Robinson felt an overpowering sense of thankfulness and thanked 'from the bottom of my heart that Supreme Power which rules and guides our destinies.' Then, in sheer exuberance, he fired off Very lights and dropped a parachute flare before returning to Suttons Farm, landing with very little petrol remaining in his fuel tank. He had been in the air for three and a half hours. Questioned by his Flight, as the BE rolled to a stop, he admitted 'Yes, I've strafed the beggar this time.' The whole Flight carried him cheering from the BE.

Left, above left, and above: The armament of BE2c 2693.

The blazing wreckage of *SL11* took some while to fall to earth and the intense fire lit up the countryside over a 60 mile radius. All the searchlights immediately went out. There was no need for their light, 'there was a new and terrifying light in the heavens.'[1] Suddenly, the huge mass exploded with a brilliant flash, illuminating the whole of London before finally crashing by the village of Cuffley in Hertfordshire, where the remains burnt for over two hours. In the wreckage the control gondola clock had stopped at 3:23 hours German time, an hour later than British Summer Time.

Robinson's was a significant victory. *SL11* was the first airship to be brought down over British soil and in full view of a war-weary populace, still stunned by the losses on the Somme. Thousands of people flocked into the streets in London and the Home Counties to cheer and watch the awesome sight of the conflagration, which had been *SL11* in its final fall to earth. The awful fear and horror of the Zeppelins seemed to have been broken for good.

Lieutenant-Colonel Henderson, commanding the Royal Flying Corps, was requested by Sir John French to make a recommendation of an award to Robinson for his destruction of *SL11*. Henderson had no doubt. His recommendation was for a Victoria Cross and the award was gazetted on 5 September 1916.

Robinson received his VC from King George V in Windsor Castle on 9 September, surely the only officer to be late for his appointment with the King. Robinson's car had broken down on his way to Windsor and the other six officers to be given awards that day had already received them by the time a distraught Robinson finally arrived at the wheel of a borrowed car. After lunch in the state dining room, Robinson was finally received by the King and given his award. The King then introduced him to the Queen, Princess Mary, and Prince Albert, before questioning the young airman on aspects of aviation and asking for details of his family.

Robinson was instantly catapulted into public fame, recognized and lionized wherever he went. He became so well known that he was forced to wear civilian clothes whenever he visited a restaurant in London, went to the theater, or one of the popular music halls. Even this disguise did not prevent him being recognized: he was saluted by policemen, people in the street stopped him and asked to shake his hand. 'The waiters, hall porters, and pages of restaurants bow and scrape, visitors turn round and stare – oh, it's too thick.' Celebrities and actresses flocked to Suttons Farm to meet the young hero of the hour.

Besides becoming the toast of the entire country, Robinson received over £4,000 in prize money from grateful donors, a considerable sum in those days. These awards were frowned upon in official circles and a regulation was quickly passed to forbid any in the future.

After a rest from flying on active service – the authorities possibly had no wish for any mishap to befall him – Robinson was posted as a Flight Commander to 48 squadron, stationed at Rendcombe in Gloucestershire, where it was working up for service in France. The squadron had been formed at Netheravon in April, under the command of Major L. Parker and was the first RFC squadron to be equipped with the new Bristol F.2A. The Bristol Fighter – as it was known – was a two-seat

1. *Zeppelin*. R. Rimmel. Conway Maritime Press. 1984.

aeroplane, powered by a 190hp Rolls Royce Falcon 1a engine, and 48 Squadron flew its new aeroplanes to France on 8 March 1917.

When its potential as a two-seat fighter was recognized, the Bristol Fighter became one of the most successful fighter aeroplanes of the Great War, but when 48 Squadron arrived in France, the concept of using its Bristols as two-seat fighters had not yet been realized and the Squadron's first patrol over the lines ended in disaster.

On 5 April 1917, Robinson led his Flight of six Bristol Fighters over the front line. Robinson had had no experience of aerial fighting in France and had ignored the advice of more experienced pilots, who had warned him not to take the Flight across the lines at the low and dangerous height of 4,000 feet. Not realizing that the Bristol could be flown in an offensive manner, by taking full advantage of the pilot's forward-firing Vickers gun, Robinson had also instilled into his Flight the necessity of keeping a tight formation, thinking that this tactic, as practiced by the crews of the FE2bs, would afford them maximum protection if attacked. One pilot later commented that the more experienced pilots in the squadron were 'not bothered' about Robinson's VC, only that he seemed to dislike and disregard any opinion, other than his own, in matters of which he had no experience.

Led by Robinson, the six Bristol Fighters left their aerodrome at La Bellevue at 10:15a.m. and crossed the lines by Arras. As they neared the area Mericourt – Douai they were attacked by five Albatros D.IIIs of *Jasta* 11, led by the *Staffelführer,* Manfred von Richthofen, who attacked the rearmost Bristol A3340, crewed by 2nd Lieutenant A.N. Leckler and Lt. H.D.K. George. Richthofen's fire wounded both pilot and observer and Leckler was forced to land at Lembras, behind the German lines. Both he and his observer were taken prisoner, George later dying of his wounds.

The Bristols' formation was now broken and they were separated, with no mutual protection. *Ltn.* Georg Simon shot down H.A. Cooper and Boldison in Bristol Fighter A3320 near Monchecourt; both were wounded and taken prisoners of war. Manfred von Richthofen, chasing after the scattered Bristols, caught H.T. Adams and his observer D.J. Stewart over Quincy, wounding Stewart and forcing Adams to land his damaged machine, No.3343, behind the German lines, where he set fire to it.

Vizefeldwebel Sebastian Festner attacked the Bristol Fighter flown by Robinson, hitting the engine of A3337 and forcing Robinson to land the damaged Bristol behind the German lines near Mericourt at 11:00a.m. Both Robinson and his observer, 2Lt. E.D. Warburton, were uninjured and were taken POW.

Of the six Bristol Fighters which had left La Bellevue, only two returned, one severely damaged. The only Bristol to escape relatively unscathed was that flown by Captain A.M. Wilkinson. Alan Wilkinson was another Flight Commander in 48 squadron and was a pilot of vast experience, having flown in France since 1915. It was his advice, with others, which Robinson had ignored with such disastrous results.

In captivity, it was not long before Robinson was identified as the pilot who had destroyed *SL11.* In the past, some reports of Robinson's treatment in various prisoner-of-war camps have suggested that he was ill treated because of this, but there is no real evidence to support this view. What is certain is that, due to his various attempts to escape – two actual escapes and recapture plus several aborted attempts – he gained a reputation as a desperate character and was guarded accordingly. In addition, as a result of his attempts to escape, he was given several spells in solitary confinement. In October 1917, he and a fellow prisoner, Baerlin, were court-martialed for attempting to bribe a prison guard. Both were found guilty and Robinson was sentenced to a month's solitary confinement in the fortress of Zorndorf, a particularly unpleasant prison.

After an unsuccessful attempt to escape from the train taking him to a new camp at Clausthal, Robinson arrived to find the camp was commanded by Heinrich Niemeyer, one of twin brothers whose harsh treatment of the prisoners under their jurisdiction was to make them both notorious. Niemeyer took a personal dislike to Robinson the moment he arrived and was determined to be rid of him. In July 1918 Heinrich Niemeyer had Robinson transferred to Holzminden, commanded by his brother Karl. Later in the year, Robinson escaped from Holzminden with Captain Stephen of 73 Squadron but they were quickly recaptured and sentenced to a period of solitary confinement. Robinson suffered from claustrophobia and this harsh treatment affected his health. In addition, Karl Niemeyer did all he could to make life unbearable for Robinson. There was no

Karl Niemeyer, 'The Beast of Holzminden.'

question of physical ill treatment, but he would be woken up at night for the guards to check on him, and bullied and harassed at every opportunity. All prisoners suspected of attempting to escape were locked into the ground floor cells of 'A' *Kaserne*. Conditions in the cells were vile, the air stale and oppressive, and there were countless flies and other vermin. Guards woke up Robinson and his fellow prisoners throughout the night, shining torches into their faces. Niemeyer's subtle cruelty failed to break Robinson's spirit, but on his release in November 1918 he was a sick man, broken in health.

On his return to England, Robinson spent Christmas at Harrow Weald; he was bent nearly double and could walk only with the aid of a stick. In 1918 the influenza plague was sweeping the world – 150,000 died in England alone – and Robinson caught the virus and was confined to bed. His health rapidly deteriorated and on 31 December 1918, William Leefe Robinson died of cardiac arrest due to influenza, tormented in his last moments by hallucinations of Niemeyer and armed guards standing by his bedside.

After Robinson's death, attempts were made to bring the Niemeyer brothers to justice for their treatment of Robinson and other officers under their charge, but the brothers both disappeared in the chaos of post-war Germany, leaving only a rumour that Karl had committed suicide.

William Leefe Robinson VC was buried on Friday 3 January 1919 at Harrow Weald Cemetery, with full military honours, aeroplanes droning overhead. His coffin was carried by fellow officers of the Royal Air Force, including Frederick Sowery, who on 24 September 1916 had emulated Robinson by bringing down a Zeppelin.

Robinson's grave is still at Harrow Weald, and the staff of The Leefe Robinson, a restaurant just across the busy road from the tiny cemetery of All Saints Church, regularly place flowers against the headstone. Every year, on Armistice day, St.Bees, Robinson's old school, sends a wreath.

Flight Sub-Lt. Reginald Alexander Warneford

When Sub-Lieutenant Rex Warneford reported to Wing Commander Arthur Longmore, the Commanding Officer of 1 Squadron, Royal Naval Air Service, based at St.Pol, France, his reputation for wildness and undisciplined behaviour had preceded him. Longmore was frank with this unruly pilot: despite his 'unsavoury reputation' he would be judged solely on his work with his new squadron. That night Warneford drove one of the squadron's Talbot tenders into a ditch and crashed it again on returning to the camp. Longmore gave the young pilot one more chance, but warned him it would be his last.

The following day, Warneford channeled his wildness into action against the enemy. Taking off in a Voisin, for a routine reconnaissance with one of Longmore's most experienced observers, John H. D'Albiac, Warneford flew along the coast at 4,000 feet. As the slow pusher biplane passed over Ostend it was heavily shelled, but despite D'Albiac's frantic signal to take them beyond the range of the anti-aircraft batteries, Warneford flew straight on to Zeebrugge, through the gauntlet of constant fire.

At St.Pol, after two and a half hours had elapsed, Longmore had given up hope of the Voisin's return, but Warneford eventually arrived, landing with only a few pints of petrol remaining in his fuel tank. D'Albiac requested that he never again be asked to fly with such a 'madman.' Warneford had chased a German aeroplane, often at rooftop height, back to its base at Ostend, firing his rifle all the way.

A braver, or perhaps more foolhardy, observer, Leading Mechanic G.E. Meddis, then volunteered to fly with Warneford and they flew many aggressive patrols together. On 17 May 1915, Warneford and Meddis had their first brush with a Zeppelin. Taking off at 3:20a.m. with orders to patrol the coast and to attack any Zeppelins they might see, they had barely cleared the aerodrome when a Zeppelin, the *LZ.39,* was seen flying towards Ostend. Warneford and Meddis closed with the giant over the town at 8,000 feet and Meddis opened fire with his rifle at a range of 1,000 feet. Meddis had time for only five shots with the new incendiary bullets before *LZ.39* immediately began to climb. Although hits were seen on the nose and amidships, the Zeppelin easily climbed away from its cumbersome attacker. Warneford then flew to Zeebrugge, where he saw a U-Boat and a small steamer leaving the harbour. Warneford dived on the submarine and threw all his small stock of hand grenades at it. The submarine appeared to be unharmed and Warneford returned to Dunkirk.

After his attack on *LZ.39,* Warneford was allocated a Morane-Saulnier L, a French-built parasol monoplane, fitted with a Lewis gun mounted in front of the pilot's cockpit. The synchronized machine gun, which would allow a forward-firing gun to be fired without damage to the propeller blades, had yet to be fully developed, but the Morane's propeller was fitted with V-shaped deflector sections which, it was hoped, would protect it from damage. Warneford's mechanics also fitted a bomb rank to the undercarriage to carry six 20lb Hales bombs, released by a lever mounted on the starboard side of the fuselage.

With the Morane, and a roving commission from Longmore, Warneford flew many lone patrols over the enemy lines attacking both aeroplanes and balloons. The Morane rarely returned without some damage and while it was being repaired Longmore placed another Morane at Warneford's disposal. This aeroplane, No.3253, lacked the machine gun of the first Morane, but bomb racks were fitted.

On the evening of 6 June 1915 an Admiralty signal reached Longmore. Two Zeppelins, the *LZ.37* and *LZ.39,* were on their way back to their base in Belgium after a raid. A third airship, *LZ.38,* had returned early with technical trouble and was back in its shed at Evère.

Sub-Lieutenant Reginald Alexander Warneford.

Longmore detailed Warneford and another Morane pilot, Rose, to attack the airships over Ghent, sending another two pilots in Farmans to bomb the enemy sheds at Evère. The plan was to either destroy the Zeppelins in the air or in their sheds after they had landed.

Warneford and Rose took off in their Moranes from the advanced landing field at Furnes at 1:00a.m. There was a heavy mist and Warneford soon lost sight of Rose, who force landed near Cassel. As Warneford neared Dixmude he suddenly saw, to the north, beyond Ostend, the unmistakable pencil shape of a Zeppelin. It was the *LZ.37,* commanded by *Oberleutnant* Otto von der Haegen, with *Obersteuermann* Alfred Muhler at the helm. Warneford attacked the Zeppelin at 1:50a.m. but he was driven off by determined fire from the enemy gunners. Another attack brought the same result and Warneford turned away. He knew the Zeppelin would have to lose height as it approached its base and he kept well to the rear, hoping to convince the airship commander that he had abandoned his attack.

LZ.37 was almost home. It had bombed Dover and Calais and was now on a general course for Brussels. Muhler later wrote, 'A weak redness was beginning to show in the morning sky. Through the gray vapours the first outlines of Ghent could be seen in the distance. We could already pick out houses and trees under us and we began to feel safe again.' It was 2:15a.m. Ten minutes later Warneford attacked. He was at 11,000 feet and he switched off his engine and dived, 'to descend on top of him.' At 7,000 feet the airship was immense under him and Warneford flew along its length, pulling his bomb release handle. As the last bomb of the six left the rack an enormous explosion sent the Morane upwards and over onto its back. *LZ.37* was almost torn in two and the burning skeleton fell across the convent of St. Elizabeth at Mont St.Amand, a district of Ghent.

Warneford struggled to regain control of the Morane – with the elementary flying training of pilots at the time he had no idea how to right the Morane. 'I had not the vaguest idea what to do, and did nothing; which was, as it happens, the right thing to do.' But he could not restart his engine and 15 minutes after destroying *LZ.37* Warneford was forced to land in enemy-held territory. A frantic inspection of the Morane revealed a broken connection between the pressure and gravity petrol tanks and it took Warneford a quarter of an hour to repair it. He could hear German cavalry searching the nearby woods and although they moved away he knew that he would certainly be captured if he could not take off. He primed the cylinders with petrol, then swung the propeller. His problem then was that, being alone, he could not keep the engine running long enough for him to scramble into the cockpit. He pulled and pushed the Morane until he was on the rise of a small but fairly steep hill. He re-swung the prop and pushed the little Morane down the hill until it began to gather momentum. When the aeroplane had reached a speed which Warneford judged would be sufficient to keep it going, he leapt into the cockpit, just as the German cavalry emerged from the woods, firing as they came.

Flying low, through mist and fog, with the engine of the Morane spluttering and missing, Warneford finally saw the sea and landed on the beach at Cape Gris Nez. As the Morane rolled to a stop it was surrounded by French troops who, despite the RNAS roundels on the fuselage and the protestations of Warneford, marched him at gunpoint – hands held high – to their commanding officer. Warneford eventually persuaded the officer – luckily he spoke English – to telephone St.Pol for confirmation of his identity. This established, Warneford was given drinks while the petrol tanks of the Morane were refilled. At 10:30a.m. an exhausted but jubilant Warneford landed at St.Pol. After making his report to Longmore he was taken to the quayside at Dunkirk, where his billet, the cross channel steamer *Empress,* was moored. He slept for eight hours.

By Tuesday, 8 June, the London newspapers had headlined the destruction of two Zeppelins – Warneford's fellow officer, Flight Sub-Lieutenant Mills, had destroyed

Morane-Saulnier L 3253 in which Warneford destroyed Zeppelin *LZ.37.*

LZ.38 in its shed at Evere. In Longmore's office at St.Pol Warneford was handed a telegram from Buckingham Palace.

I most heartily congratulate you upon your splendid achievement of yesterday in which you single-handed destroyed an enemy Zeppelin. I have much pleasure in conferring upon you the Victoria Cross for this gallant act.

George RI.

Not unnaturally, Warneford was overwhelmed by the award. He would be needed in London, but until then Longmore was ordered to keep him out of trouble while the propaganda aspect of his victory was exploited. Longmore grounded the new hero but the French authorities then stepped in. They wished Warneford to go to Paris to receive the Knights Cross of the Legion of Honour and attend various dinners and celebrations. Why Longmore consented to let Warneford go to Paris seems strange, in view of Warneford's known propensity for wild behaviour, but he detailed Flight Lieutenant Michael Marsden to accompany Warneford to the French capital.

On 12 June Warneford, with Marsden, went to the *Ministére de la Marine* where the French Minister of War, Monsieur Alexandre Millerand, pinned his own *Chevalier of the Légion d' Honneur* on Warneford's tunic.

Wherever he went in Paris, Warneford was fêted and honoured. Crowds cheered him in the streets and on more than one occasion it was necessary to provide him with an escort of gendarmes.

On the afternoon of 17 June, Warneford left Paris for the aerodrome at Buc. His orders were to fly an acceptance test of a replacement aeroplane for the squadron, a Henry Farman F.27. At noon, over lunch, he had given an interview to Henry Beech Needham, an American newspaper man, and it was arranged that he would accompany Warneford to the aerodrome for a flight in the Farman. After lunch, Warneford returned to the Ritz, where he was staying. In the lift he met an officer in the Navy, a Lieutenant R.F. Lee-Dillon, who was on leave in Paris with his wife. Lee-Dillon introduced himself. Warneford asked Lee-Dillon if he would like to have a flight that afternoon as he would be leaving shortly for Buc to test a Farman. Lee-Dillon had never flown, was far from happy at the thought, and politely declined, but when he returned to his room his wife persuaded him that he would be foolish to turn down the opportunity of a flight with such a famous pilot. Lee-Dillon ran downstairs and told Warneford that he had changed his mind and he, his wife, Warneford and another RNAS officer – probably Marsden – left in a RNAS Fiat driven by a French military driver. After a stop at RNAS Headquarters in the Avenue Montaigne, where the other RNAS officer left them, his place being taken by Needham, the party drove to Buc. On arrival at Buc, the Farman F 27 was wheeled out. Lee-Dillon, still extremely nervous, climbed aboard. Within five minutes the test flight was over; Lee-Dillon was both relieved and elated. His wife was to be the next passenger, but Lee-Dillon later wrote 'for some unknown reason I began to make objections to her going.' Warneford suggested that as time was passing he would take Needham up while Lee-Dillon and his wife 'argued it out.' Needham, as fearful of the flight as Lee-Dillon had been, climbed aboard and Warneford took off.

Lee-Dillon watched the Farman take off. 'I saw them go off beautifully, no suspicion of anything wrong. They rose to a height of about three hundred feet and turned to the left. Then the machine dived about 50 feet, fairly steeply. He straightened out and at once began what seemed to be another dive. To my horror, Warneford shot straight out like a dart, diving into a field of wheat. At once the wings crumpled up, the whole machine burst into flames and crashed about three quarters of a mile from where we were standing.'

Lee-Dillon ran to the scene, 'till I thought my chest would burst.' Ordering soldiers who had run from a nearby building to hold the roads, and to commandeer the first car they saw, 'no matter whose it was,' Lee-Dillon began to walk through the tall wheat looking for Warneford and Needham. Needham they found almost at once and it was obvious that he was beyond help. It took some time to find Warneford, '...he was in a terrible

Above and right: Morane-Saulnier L 3253 in which Warneford destroyed Zeppelin *LZ.37.*

state, but still alive. I felt under his tunic, putting my hand on his heart. I remember being moved almost to tears by the sight of his pathetic Legion of Honour, received but that morning (sic) with such joy, now damaged and actually sticking into his chest.'

The soldiers had obtained a civilian car. Lee-Dillon had Warneford placed in it and told the driver 'to rush at full speed to the nearest hospital, which was Versailles.' Lee-Dillon followed in another car.

Warneford was taken to the British Military Hospital in the Versailles Trianon Palace Hotel. Marsden, who had heard of the tragedy, arrived at the hospital as the car arrived with Warneford, followed by Warneford's friend, the Baroness de Laroche.

'As we stood helplessly looking, he gave a little sigh and opened his eyes. For less than a second he looked straight at me, almost as though he was going to make a sign that he knew us. Then his head turned away and before the orderlies had wheeled him out of our sight he died.'

Within days, speculation as to the cause of the tragedy was rife. Sabotage, foolhardiness, and even alcoholism were mentioned. The latter seems particularly unlikely in view of Lee-Dillon's nervousness. He would hardly have flown with a pilot who was the worse for drink.

On Monday 21 June 1915, Warneford's body left France en route for England. He lay overnight in the chapel of the Brompton Cemetery. The next morning thousands of people arrived to pay their last respects to the first airman to bring down one of the dreaded Zeppelins. When the police were finally forced to close the iron gates of the cemetery, thousands lined the paths to the grave site.

Reginald Alexander John Warneford was buried in a fern and laurel-lined grave just after 4:00p.m. Three volleys shattered the silence, the buglers sounded the Last Post and men of the Naval Division presented arms. Thousands of people filed past the grave before it was filled.

On 11 July Lord Derby unveiled a memorial stone, subscribed for by readers of The Daily Express, a fund which was originally set up to make a presentation to the young airman. The stone stands to this day in London's Brompton Cemetery.

Captain Ferdinand Maurice Felix West

On 4:20a.m. on August 8 1918, more than 2,000 allied guns opened fire on the German positions along the 20-mile front between Morlancourt and La Neuville. Following hard on the heels of the artillery barrage, Australian and Canadian troops, supported by tanks, advanced through the heavy ground mist; the Battle of Amiens had begun.

The German forces were taken completely by surprise; aided by the mist, the Australian and Canadian troops had in places advanced between seven and eight miles by the end of the day's fighting. General Ludendorff later described 8th August 1918 as 'the black day' of the German army during the war and the official German monograph detailing the battle used the word 'catastrophe' in its title.

For the squadrons of the Royal Air Force it was a day of heavy and complex fighting, with many casualties. Historian Trevor Henshaw commented, 'There is no other day on which casualties remotely approach those sustained this day.'[1]

By this stage of the war, after the Battle of Cambrai in 1917 and the German offensives in March and April 1918, the value of aeroplanes in support of ground forces was now fully understood and the tasks given to the RAF during the battle reflected this. The squadrons were to carry out contact patrols, attack enemy troops and positions from low level, and generally support the infantry and tanks in every way possible. In addition, the usual work of the RAF was not to be neglected; artillery spotting, bombing, and fighter patrols to destroy enemy fighter and reconnaissance aeroplanes were still to be flown.

On 8th August, Major Trafford Leigh-Mallory's 8 Squadron was stationed at Vignacourt, equipped with the Armstrong Whitworth FK8, a two-seat reconnaissance bomber. The Squadron was under the command of 15th Wing and had been attached to the Tank Corps since 1 July for experiments in closer cooperation with tanks. Its orders on the morning of August 8 were to support the tanks by the dropping of messages and seeking out enemy anti-tank guns.

One of 8 Squadron's most experienced pilots was in action early on the morning of the battle. Flying through the heavy mist which covered the battlefield during the early part of the day, Captain F.M.F. West and his observer, Lieutenant J.A.G. Haslam, obtained important information on the extent of the Allied advance, but on returning to Vignacourt found the aerodrome completely obscured by thick fog. Rockets were fired to assist West to land but the FK8 wiped off its undercarriage in landing and West and Haslam were cut and bruised in the crash. It was the beginning of a dramatic three days for West; three days which would culminate in the award of a Victoria Cross.

Ferdinand Maurice Felix West was born in Paddington, London, on 29 January 1896, the only son of Lieutenant Francis West and the Comtesse Clemence de la Garde de Saignes. The Wests were a naval and military family; Ferdinand's grandfather was Admiral of the Fleet, Sir John West, and his father was commissioned in the East Lancashire Regiment. In 1902, serving in the Boer War, Francis West was killed in action and his widow left England and took her young son to Milan, where he was educated.

In 1914 West was studying law at the University of Genoa and was on holiday in Switzerland when war was declared. Making his way to England, West initially enlisted in the Royal Army Medical Corps as a private, but in May 1915 he was commissioned in the Royal Munster Fusiliers and was in France by November. West had served four months in the trenches when he was given a flight in a Morane of 3 Squadron and he immediately decided to transfer to the Royal Flying Corps. His application having been approved, West returned to England in March 1916 and began to train as an observer, returning to France and posted to 3 Squadron at the end of April. While flying as an observer with 3 squadron, West determined to become a pilot and he put in as many flying hours as he could before applying for pilot training. By October 1917, West had accumulated 225 hours as an observer and he left 3 Squadron in October 1917 to begin his pilot training. He flew solo on 15 November and had a total of 60 hours by the end of December 1917, when he was posted to 8 Squadron, and crewed with Lieutenant J.A.G. Haslam.

West and Haslam flew throughout the German offensives of March and April of 1918. On 22 April, during the Battle of the Lys, they set out in FK8 C8445 to bomb a concentration of German transport well behind the lines. Coming down to low level, they bombed and machine-gunned until they had exhausted their ammunition. As they flew across the front line on the way back to their base at Vignacourt, anti-aircraft fire hit the FK8 in the engine and shot away an aileron. West managed to bring the damaged aeroplane down in a forced landing only a hundred yards behind the British front-line trenches. Both West and Haslam were awarded a Military Cross.

During the next two months, West had many combats with enemy fighters, claiming two Pfalz D.IIIs on 18 June, one to himself and one to his observer in the fight, Lt. D.R. Sharman. The next day he was promoted Captain and given command of a Flight.

The cuts and bruises sustained in the crash landing on 8 August did not keep West and Haslam on the ground. They were up by 10:30a.m. the next day on contact patrol with tanks. Seeing four surrounded by enemy troops, they went down to low level and attacked the infantry, but the engine of the FK8 was hit by return fire and put out of action. West turned for the front line and managed to put FK8 D5138 down in a field near Caix.

1. *The Sky Their Battlefield.* Trevor Henshaw. Grub St. 1995.

Ferdinand Maurice Felix West.

Both West and Haslam were unhurt. They were one of four FK8s lost that day by 8 Squadron.

On 10 August, West and Haslam took off again, this time in FK8 C8602. Their orders were to support tanks moving up to Roye. North-east of the town, through a gap in the low cloud, West saw a great number of enemy infantry concentrated around a large wood, somewhere in the vicinity of Ham and Hombleux. West went down through the cloud and, under heavy fire from the ground, flew over the troops in an attempt to estimate their number. This done, West again climbed into the safety of the cloud. However, he was still unsure of the exact location of the wood and he again made for the gap in the cloud so that he could return and attempt to fix its position. As he did so he was attacked by a formation of enemy fighters. One burst smashed the wireless transmitter and wounded West in the right foot. Despite his wound, West again dived through the cloud and satisfied himself of the exact position of the troops. Anxious to get the information back to base, West then made for home, but he again came under attack by five or more enemy fighters. West received five bullet wounds in the leg, one of the bullets, an explosive round, smashing his left femur and cutting the femoral artery. The now useless leg 'flopped around' and West was unable to control the FK8. He almost lost consciousness and the FK8 was diving out of control. West grabbed his leg, wedged it out of the way of the rudder controls and managed to pull out just above the trees. One of the enemy fighters was persistent and followed the FK8 down, wounding Haslam with its fire.

West was bleeding badly now, but with his left hand he wound the leg of his shorts around his leg into a make-shift tourniquet. He was drifting in and out of consciousness and he knew he would not be able to hold out long enough to return to Vagnacourt. Looking round for a safe area to land, he saw a suitable field. Lifting his now useless left leg from the rudder bar, he glided in to land, still being fired at by the enemy fighter.

West fainted as the damaged FK8 rolled to a stop, but Canadian troops ripped away the fabric from the fuselage side and lifted him gently out. West and Haslam were at once put into an ambulance – Haslam later said 'We'd landed nearly on top of two motor ambulances' – and taken to the nearest casualty clearing station. West was in great pain from his wounds and only conscious for brief moments, but he insisted that a message be got to the squadron asking for an officer to be sent so that he could report the position and number of the enemy troop concentration. Finally, the recording officer of 8 Squadron arrived and West was at last able to make his report before he lapsed again, this time into deep unconsciousness. He, along with Haslam, was then taken to Rouen. West later wrote, 'The journey to the CCS was awful but the worst was the journey down to Rouen – they kept us lying in the open, no shelter.' At Rouen, West's severely injured leg was amputated.

As soon as his condition allowed, West was evacuated to England and taken to the London Hospital in Whitechapel; he was still there on 18 November 1918 when he was told that he had been awarded a Victoria Cross for his action on 10 August. The following month he was discharged from hospital, given a wooden leg and £250 compensation from the Air Ministry. West received his VC at Buckingham Palace on 1 March 1919.

The war over, West decided to return to his law studies, but he then met Marcel Desoutter. Desoutter had also lost a leg and, being an engineer, had designed an artificial leg which enabled the wearer to dispense with sticks or crutches. West purchased a leg and, once he had gained expertise in its use, he went to see Trenchard, then Chief of the Air Staff, with the hope that he could rejoin the RAF. West had a sympathetic hearing from Trenchard. As a result he was given a temporary commission and posted in April 1919 to the Foreign Office as a liaison officer. In August 1919, West's commission was made permanent – his observer, Alec Haslam and his old CO in 8 squadron, Leigh-Mallory, were also on the short list – and he stayed in the RAF throughout the inter-war years, commanding 4 Squadron

West (on right) and his observer, Lt. J.A.G. Haslam, with Armstrong Whitworth FK8 C8594. This aeroplane has been wrongly given in the past as the FK8 in which West won his VC, but during the research for this book the casualty report for the action on 10 August 1918 showed that the FK8 in which West won his VC was C8602. West and Haslam regularly flew C8594 until West crashed it on 24th April 1918, several months before his VC action.

in the mid-thirties. In 1940 he was an Air Commodore and in June was stationed in the British Legation in Berne, Switzerland. While in Berne, West was engaged in counter-intelligence and espionage work – at one time the Gestapo put a price on his head – setting up an escape route for aircrew who would otherwise have been interned in Switzerland for the duration of the war.

At the end of the Second World War, West was given a CBE and he finally retired from the RAF in March 1946. Freddie West died in July 1988, the last surviving British holder of the VC in World War One. He was survived by his observer on that day in August 1918 by 17 months, Alec Haslam dying in December 1990.

Aircraft flown in Victoria Cross Actions

Some of the color references used are given here.

Ball, Albert — Nieuport 17 A213

A201 was the first Nieuport received by No.60 Squadron from No.11 Squadron, flown in by Ball around 16th August and was flown by him until going on leave 1 Sept. 1916. On returning Ball took over A213 which he used until being posted to H.E. on 1st October 1916. A213 was used by several other pilots until 6th MArch 1917 when Major Graves was shot down in flames in it and KIA at 12:30 hours north of Riviere.

Research by Les Rogers and Squadron Ldr. Joe Warne shows that the original Nieuports bought from the French before November 1916 were in French olive drab topsides with gray or aluminum cowls. Silver aircraft started being delivered during November 1916.

Squadron codes on 60 Squadron Nieuports started during autumn 1916. The Flight letter was placed in front of the fuselage roundel while the individual flight number was behind the roundel. These were also carried on the rear fuselage decking. Although no photo of A213 has surfaced. it is assumed that Ball would have carried 'A1.' In addition Ball also had a red *cone de penetration* fitted to his aircraft.

Les Rogers Correspondence
Cross & Cockade (GB) 12/1
JM Bruce: *Aeroplanes of the RFC*

Barker, William George — Sopwith Snipe E8102

When Barker went back to the Western Front on a 'refresher course' he took with him what was to become one of the most famous aircraft of the First World War – Sopwith Snipe E8102. E8102 was finished in PC10, with grey cowl and plywood panels. On the aft fuselage Barker had painted 5 white bands, in addition to which he added his mascot which had previously been on Sopwith Camel B6313. This consisted of a sheet metal red devil thumbing its nose and was placed on the muzzle of the right-hand Vickers. This is depicted as done in a sketch

by Captain Ronald Sykes of No.201 Squadron in 1919.
Datafile # 46
Canadian War Museum Pamphlet
Cross & Cockade (GB) 2/2

Beauchamp-Proctor, Anthony W. RAF SE5a D6856

The Wolseley-built D6856 was finished in overall PC10. No 84 Squadron's marking consisted of two inward-sloping white bars on either side of the fuselage roundel. Beauchamp-Proctor's 'C' Flight featured blue and white wheel covers. Unusual for a Scout squadron, No.84's aircraft had no separate flight letter/number.
Aerodrome Modeler 4/3

Bell-Davies, Richard NIeuport 10 3172

3172 is in plain finish with modified RNAS roundels. When the decision to follow the RFC practice of blue/white/red roundels was made, aircraft with the existing RNAS roundel of red/white were allowed to add a blue centre, in effect making 'French' roundels on British aircraft. In an attempt to arm the aircraft a Lewis gun was fitted and angled up through center section.
Sturtivant/Page: *RN Aircraft Serials: and Units 1911–1920:* 90
Over The Front 9/1
Alex Revell Correspondence

Bishop, William Avery Nieuport 17 B1566

This aircraft was in overall aluminum finish with French-proportioned roundels on wings. Blue cowl. 'C5' on fuselage sides and rear fuselage decking, with 'C' foremost on both sides. Note that serial is outlined in red on the white portion of the rudder stripe and white on the red and blue areas. A period is used to separate the 'B' and '1,' not an accent as seen in most other sources.
Alex Revell correspondence
Baker: *Billy Bishop – The Man and his Aircraft*
Cross & Cockade (GB) 12/1

Hawker, Lanoe George Bristol Scout 'C' 1611

1611 was plain finished overall with varnished plywood panels and metalwork left natural. No roundels on upper wing, but with roundels on bottom surface of lower wing. Lewis gun fitted at 45° angle to line of flight.
Cross & Cockade (Int) 19/2
Datafile No.44

Insall, Gilbert Vickers FB5 5074

Overall plain fabric with natural metal nacelle.
Datafile No.56

Jerrard, Alan Sopwith Camel B5648

B5648 was finished as per Ruston Proctor practice of overall PC10 with cowl and side panels painted the same. No.66 Squadron's marking consisted of a vertical white bar forward of the fuselage roundel and a horizontal stripe from the roundel to the leading edge of the tailplane. B5648 carried the flight letter 'E' on fuselage sides and was repeated on the upper wing centre section. In addition the roundels of the upper wing had their white portions roughly smudged out using either paint or dirt. Ruston Proctor put their serials on the fin in black, later replaced by grey although many units (No.66 Squadron for one) gave the serial a thin white outline.
Bowyer: *Camel: King of Combat*
Sturtivant/Page: *The Camel File*

Liddell, John RAF RE5 2457

This is one that, unfortunately, some speculation must creep into. All we can be sure of is the pattern on the port elevator, starboard fin, and starboard lower wing and parts of the top wing. Various portions of the fuselage and the starboard tailplane can also be made out with considerable headache the inevitable result. The serial on the rudder is in white and blue. Camouflage colours are unknown, but dark green is a reasonable assumption.
Cross & Cockade (Int.) 25/3

Mannock, Edward RAF SE5a D278

No.74 Squadron featured a white horizontal bar. As 'A' Flight commander, Mannock carried the letter 'A' on his aircraft, the beginning of which can be made out in the familiar photo of him in the cockpit of D276.
Alex Revell correspondence
Over The Front 5/1

McCudden, James Thomas Byford RAF SE5a B4863

B4863 had the 56 Squadron 18″ white band on the aft fuselage, flight letter 'G,' short exhaust, modified cockpit. Top and bottom views are also included showing inboard roundels with no white outer rings.
Revell: *Aces and Aeroplanes No.3*
Revell: *High In the Empty Blue*

McLeod, Alan Armstrong Whitworth FK8 B5773

As the RFC was ordered to remove squadron markings with the German offensive in March 1918, it is assumed that No.2 Squadron complied and B5773 is depicted with no zig-zag, although McLeod's earlier B5782 was photographed by him with the zig-zag in Dec. 1917. B5773 was fitted with the original form of oleo undercarriage and radiator and a rounded nose.
S.K. Taylor photograph of B5782
Alex Revell correspondence

McNamara, Frank Martinsyde G.100 7486

7486 was probably finished in overall plain as shown on the similiar 7488.
Cross & Cockade (GB) 4/4
Schaedel: *Australian Flying Corps*

Mottershead, Thomas RAF FE2d A39

Overall PC10. Serial on angle on rudder as per others in series A1 to A40. Roundels were inboard of the ailerons.
Bruce: *Aeroplanes of the RFC*
Windsock 10/6

Rees, Lionel — deHavilland DH2 6015

Photos of No.32 Squadron prior to leaving for France show them to be finished in PC10 and grey. 6015 is depicted as an aircraft of 'A' flight with white wheel covers.

Cross & Cockade (Int) 21/3
Datafile No.48
S.K. Taylor correspondence

Rhodes-Moorhouse, William — RAF BE2b 687

Plain finish, Union Jack on rudder. Previously the Jack had also been displayed on the bottom surface of the lower wings; however. this had been removed by early 1915.

Alex Revell correspondence
Les Rogers correspondence
Profile No.133
Squadron/Signal *BE2 in Action*

Robinson, William Leefe — RAF BE2c 2693

Plain finished with PC10 cowl and nose panels. No underwing roundels. Inboard portion of underside starboard bottom wing in PC10. Holt flare fittings under bottom wing with a black silk panel to reflect the flare upon landing. Lewis gun on Strange mounting.

Windsock 5/2
Rimell: *Zeppelin*
Rimell: *The Airship VC*
Cross & Cockade (GB) 19/2
Datafile No.42

Warneford, Reginald A J — Morane-Saulnier Type L 3253

Long nose. Plain overall with French roundels on wings and British fuselage roundel. White cowl. Black serial with white drop shadow.

Datafile No.16
Sturtivant/Page: *RN Aircraft Serials and Units: 1911–1919*

West, Ferdinand — Armstrong Whitworth FK8 C8602

Aircraft in which he was shot down and won the VC on 12 August 1918 (not C8594). No 8 Squadron RAF. Overall PC10 with plain wheel covers and no markings other than serial in photo. C8602 was a later production aircraft fitted with flank radiators, round nose, and Bristol F2b undercarriage.

S.K. Taylor photo of C8594
Alex Revell correspondence

Other Aircraft Used by VC Winners

Barker, William G — Sopwith Camel B6313

B6313 went through many changes to its appearance during its year of combat flying. Initially it had No.28 Squadron's white square and wheel covers and flight number of '1' with a small 'C,' possibly changed to 'N.' When Barker went to No.66 Squadron, B6313 most likely had their markings, but no photo has been found. Upon going to No.139 Squadron it had two white bands behind the roundel, changing to w/b/w/b/w/b/w and later expanding to seven white and six black extending from the roundel to the tailplane. B6313 also had an arrow through the heart on its fin. At some point this was repainted as shown with the arrow facing both fore and aft at different times. Barker also kept a tally of his victories on the front port interplane strut, eventually reaching 37 of them. Here B6313 is depicted in its ultimate appearance with No.139 Squadron RAF in September 1918.

Bowyer: *Camel: King of Combat*

Acknowledgements

I would like to acknowledge the help so freely given by friends and members of Cross and Cockade International in providing photographs for this book.

Jack Bruce and Stuart Leslie, from their magnificent collection; Frank Cheesman; Norman Franks; Barry Gray; Colin Huston; Alex Imrie; Paul Leaman; Les Rogers; Phil Jarrett; Ray Sturtivant. My thanks to them all.

Alex Revell

Credits

Color illustrations and references: Bob Pearson
Book and cover design and layout: Jack Herris
Scanning and digital editing: Jack Herris

ISBN 1-891268-00-7

For the latest information about new products from Flying Machines Press, visit our web site at:
www.flying-machines.com

Flying Machines Press
35 Chelsea Street
Stratford, CT 06497, USA
Tel: 203-378-9344
Fax: 203-380-1849

More Quality Books from Flying Machines Press

French Aircraft of the First World War

By Davilla and Soltan. This landmark book — the first to cover this important topic — is presented with the quality, photos, scale drawings, and color illustrations you have come to expect from Flying Machines Press.
(ISBN 0-9637110-4-0) **$96 in US, $99 International**

- More than 650 over-size (9" by 12") pages covering all 400 French WWI aircraft types built.
- 983 photos, many rare and previously unpublished.
- 185 three-view aircraft drawings, most to standard modeler's scales of 1/72 (130 drawings, including essentially all operational types) and 1/144 (50 drawings), many never before available. (5 drawings not to scale, 34 drawings on fold-out pages.)
- 25 pages of exceptional color illustrations by Alan Durkota showing 61 aircraft, including four views of a dozen aircraft types in 5-color camouflage. Two full-page color paintings and 8 color photos of museum aircraft bring the total to 30 pages of color.
- Approximately 300,000 words of text, including extensive operational details and order of battle information never before published.

The Imperial Russian Air Service

By Durkota, Darcey, and Kulikov. The first comprehensive coverage of the Russian Air Service in WWI.
(ISBN 0-9637110-2-4) **$86 in US, $89 International**

- Hard bound, 560 oversize (9" by 12") pages.
- 615 rare photos, 20 drawings and maps.
- Scale drawings of 40 aircraft (30 @ 1/48 and 10 @ 1/72 scale).
- 48 pages of color illustrating 85 planes (including many aces' planes) plus uniforms and medals.
- Detailed biography of each Russian ace.
- Detailed biography of each foreign ace in Russia.
- Famous Russian designers Sikorsky, Grigorovich).
- Key Russian manufacturers (Anatra, Dux, Lebed).
- Overview of army and naval aviation and EVK.
- Distinguished Russian aviators, first women combat pilots, etc.
- Appendices: lighter-than-air, Russian awards, pilot victory lists, aircraft flown by aces.

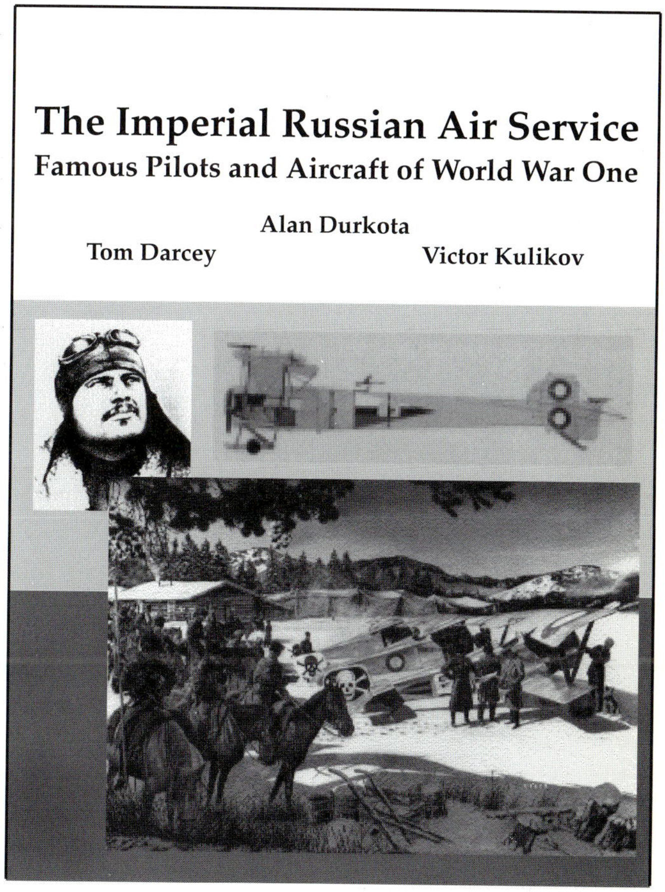

High in the Empty Blue

(ISBN 0-9637110-3-2) Alex Revell's definitive history of Britain's premier WWI fighter squadron, 56 Squadron RFC/RAF. History of every air combat, etc. **$54 US, $58 Int'l**

- 430 photos, 30 documents & drawings.
- 8 pages of color showing 17 aircraft.
- SE5 & SE5a 3-views (1/48 & 1/72 scales).
- 21 appendices list all squadron personnel, aircraft, aerodromes, victories, casualties, 'Greentail,' etc.
- Hard-bound, 450 pages (8.5" by 11").

Austro-Hungarian Army Aircraft of World War One

(ISBN 0-9637110-0-8) by Peter M. Grosz, G. Haddow, P. Schiemer. The definitive reference on these exotic aircraft. **$91 US, $95 Int'l**

- 903 photographs and 25 drawings.
- 26 pages of color illstrating 56 aircraft (all new profiles complementing the color in *Air Aces of the Austro-Hungarian Empire*).
- 102 three-views (62 in 1/48, 40 in 1/72).
- Detailed descriptions of aircraft built by the ten Austro-Hungarian manufacturers.
- Helicopter development 1914–1918.
- Full coverage of German and foreign aircraft purchased by Austria-Hungary.
- Twelve appendices, including production schedules into 1919, armament, photo single-seaters, projects, etc.
- Hard-bound, 570 pages (10" by 12").

Air Aces of the Austro-Hungarian Empire 1914–1918

(ISBN 0-9637110-1-6) by Martin O'Connor. The most detailed work on these courageous airmen. **$54 US, $58 International**

- The full story of all 49 Austro-Hungarian aces.
- 240 rare photos.
- 16 color pages showing 50 aces' aircraft.
- Hard-bound, 336 pages (8.5" by 11").